T0140065

Springer Theses

Recognizing Outstanding Ph.D. Research

Aims and Scope

The series "Springer Theses" brings together a selection of the very best Ph.D. theses from around the world and across the physical sciences. Nominated and endorsed by two recognized specialists, each published volume has been selected for its scientific excellence and the high impact of its contents for the pertinent field of research. For greater accessibility to non-specialists, the published versions include an extended introduction, as well as a foreword by the student's supervisor explaining the special relevance of the work for the field. As a whole, the series will provide a valuable resource both for newcomers to the research fields described, and for other scientists seeking detailed background information on special questions. Finally, it provides an accredited documentation of the valuable contributions made by today's younger generation of scientists.

Theses are accepted into the series by invited nomination only and must fulfill all of the following criteria

- They must be written in good English.
- The topic should fall within the confines of Chemistry, Physics, Earth Sciences, Engineering and related interdisciplinary fields such as Materials, Nanoscience, Chemical Engineering, Complex Systems and Biophysics.
- The work reported in the thesis must represent a significant scientific advance.
- If the thesis includes previously published material, permission to reproduce this must be gained from the respective copyright holder.
- They must have been examined and passed during the 12 months prior to nomination.
- Each thesis should include a foreword by the supervisor outlining the significance of its content.
- The theses should have a clearly defined structure including an introduction accessible to scientists not expert in that particular field.

More information about this series at http://www.springer.com/series/8790

Antonio Artuñedo

Decision-making Strategies for Automated Driving in Urban Environments

Doctoral Thesis accepted by
Universidad Politécnica de Madrid, Spain

Author
Dr. Antonio Artuñedo
Centre for Automation
and Robotics (CSIC-UPM)
Spanish National Research Council
Madrid, Spain

Supervisors
Dr. Jorge Villagrá
Centre for Automation
and Robotics (CSIC-UPM)
Spanish National Research Council
Madrid, Spain

Dr. Rodolfo Haber
Centre for Automation
and Robotics (CSIC-UPM)
Spanish National Research Council
Madrid, Spain

ISSN 2190-5053 ISSN 2190-5061 (electronic)
Springer Theses
ISBN 978-3-030-45907-9 ISBN 978-3-030-45905-5 (eBook)
https://doi.org/10.1007/978-3-030-45905-5

This Springer imprint is published by the registered company Springer Nature Switzerland AG
The registered company address is: Gewerbestrasse 11, 6330 Cham, Switzerland

The scientists of today think deeply instead of clearly. One must be sane to think clearly, but one can think deeply and be quite insane.

Nikola Tesla

Supervisor's Foreword

Intelligent vehicles technology is advancing at a vertiginous pace. However, the complexity behind some highly uncertain and dynamic urban driving scenarios is preventing a massive deployment of autonomy-enabled mobility solutions. To cope with this complex problem, a breakthrough is required in embedded decision-making approaches, so that the influence of world modelling, localization and mapping uncertainty can be appropriately taken into consideration.

The Ph.D. thesis entitled *Decision-making Strategies for Automated Driving in Urban Environments*, authored by Antonio Artuñedo, provides ground-breaking solutions in this research field. It covers and contributes to some of the highly specialized disciplines involved in automated driving for complex and uncertain environments (e.g., mapping, control, planning, etc.). The thesis contributions not only cover a significantly wide scope and are structured around a novel decision-making architecture, but they are also showcased with real in-vehicle experiments, which inherently entail a huge integration effort of significance value for practitioners. Furthermore, this thesis starts by presenting a detailed discussion of the state-of-the-art in automated vehicles, which allows to easily contextualize the different contributions of the work.

Another relevant aspect of this work is that each of the contributions is supported by both theoretical and extensive experimental validation. This has allowed such results to be presented and published in relevant conferences and journals. A good example of that relevance is the successful demonstration conducted at the IEEE/RSJ International Conference on Intelligent Robots and Systems (IROS'18), which is one of the most relevant venues for the robotics community around the world.

The designed and validated decision-making strategy paves the way for additional research activity and potential technology transfer, thus aiming to contribute to the deployment of novel urban mobility paradigms, where autonomous vehicles will play a crucial role.

Madrid, Spain
January 2020

Dr. Jorge Villagrá

Parts of this thesis have been published in the following articles:

Journal Papers

- **Artuñedo, A.**, Godoy, J., & Villagra, J. (2018). "A Primitive Comparison for Traffic-Free Path Planning", IEEE Access, 6, 28801–28817. https://doi.org/10.1109/ACCESS.2018.2839884
- **Artuñedo, A.**, Villagra, J., & Godoy, J. (2019). "Real-time motion planning approach for automated driving in urban environments", IEEE Access, 7, 180039–180053. https://doi.org/10.1109/ACCESS.2019.2959432
- **Artuñedo, A.**, Toro, R. del, & Haber, R. (2017). "Consensus-Based Cooperative Control Based on Pollution Sensing and Traffic Information for Urban Traffic Networks", Sensors, 17(5), 953. https://doi.org/10.3390/s17050953
- Godoy, J., **Artuñedo, A.**, & Villagra, J. (2019). "Self-Generated OSM-Based Driving Corridors", IEEE Access, 7, 20113–20125. https://doi.org/10.1109/ACCESS.2019.2897348
- Castaño, F., Beruvides, G., Haber, R., & **Artuñedo, A.** (2017). "Obstacle Recognition Based on Machine Learning for On-Chip LiDAR Sensors in a Cyber-Physical System", Sensors, 17(9), 2109. https://doi.org/10.3390/s17092109

Book Chapters

- Villagra, J., Acosta, L., **Artuñedo, A.**, Blanco, R., Clavijo, M., Fernández, C., Godoy, J., Haber, R., Jiménez F., Martínez, C., Naranjo, J. E., Navarro, P. J., Paúl, A. & Sánchez, F. (2018). "Automated Driving". In "Intelligent Vehicles, Enabling Technologies and Future Developments" (pp. 275–342). Elsevier. https://doi.org/10.1016/B978-0-12-812800-8.00008-4
- Sánchez-Medina, J.J., Arnay, R., **Artuñedo, A.**, Campos-Cordobés, S., & Villagra, J. (2018). "Simulation Tools—Traffic Simulation". In "Intelligent Vehicles, Enabling Technologies and Future Developments" (pp. 395–436). Elsevier. https://doi.org/10.1016/B978-0-12-812800-8.00010-2

Conferences

- **Artuñedo, A.**, Godoy, J., & Villagra, J. (2019). "A decision-making architecture for automated driving without detailed prior maps" In 2019 IEEE Intelligent Vehicles Symposium (IV) (pp. 1645–1652). IEEE. https://doi.org/10.1109/IVS.2019.8814070
- **Artuñedo, A.**, Godoy, J., & Villagra, J. (2017). "Smooth path planning for urban autonomous driving using OpenStreetMaps", In 2017 IEEE Intelligent Vehicles Symposium (IV) (pp. 837–842). IEEE. https://doi.org/10.1109/IVS.2017.7995820

- **Artuñedo, A.**, Godoy, J., Haber, R., Villagra, J., & Toro, R. M. del. (2015). "Advanced Co-simulation Framework for Cooperative Maneuvers Among Vehicles", In 2015 IEEE 18th International Conference on Intelligent Transportation Systems (pp. 1436–1441). IEEE. https://doi.org/10.1109/ITSC.2015.235
- **Artuñedo, A.**, Godoy, J., & Villagra, J. (2017). "A comparison of local path-planning interpolation methods for autonomous driving in urban environments," In Industriales Research Meeting 2017 (p. 147). Madrid: ETSII, UPM. Retrieved from http://oa.upm.es/46090/
- **Artuñedo, A.**, Haber, R., & Matía, F. (2016). "A co-simulation environment for cooperative maneuvers among vehicles", In Industriales Research Meeting 2016 (p. 219). Madrid: ETSII, UPM. Retrieved from http://oa.upm.es/40073/
- **Artuñedo, A.**, Villagra, J., & and Haber, R. (2018). "Adaptive and cooperative decisionmaking strategies for autonomous driving in urban environments", In II Doctorate Symposium—Universidad Politécnica de Madrid". Madrid. Retrieved from https://eventos.upm.es/file_manager/getFile/25023.html
- Villagra, J., Perarnau, M., Godoy, J., & **Artuñedo, A.** (2018). "Validación de una estrategia para la estimación del riesgo en intersecciones con vehículos conectados", In Actas de las XXXIX Jornadas de Automática, Badajoz, Universidad de Extremadura. http://eii.unex.es/ja2018/actas/JA2018_030.pdf
- **Artuñedo, A.**, del Toro, R. M., & Haber, R. (2016). "Consensus-Based Cooperative Control Approach Applied to Urban Traffic Networks", Proceedings, 1(2), 29. https://doi.org/10.3390/ecsa-3-e008
- **Artuñedo, A.**, del Toro, R. M., & Haber, R. (2016). "Sistema de control cooperativo aplicado a una red de tráfico urbano", In Actas de las XXXVII Jornadas de Automática (pp. 558–565). Madrid: Comité Español de Automática (CEA-IFAC). Retrieved from http://ja2016.uned.es/assets/files/ActasJA2016.pdf
- **Artuñedo, A.**, Godoy, J., & Haber, R. (2016). "Entorno avanzado de co-simulación para maniobras cooperativas entre vehículos", In Actas de las XXXVII Jornadas de Automática (pp. 704–709). Madrid: Comité Español de Automática (CEA-IFAC). Retrieved from http://ja2016.uned.es/assets/files/ActasJA2016.pdf
- Godoy, J., **Artuñedo, A.**, Haber, R., & González, C. (2015). "Conducción autónoma y cooperativa—El programa Autopia en España", In XV Congreso Español sobre Sistemas Inteligentes de Transporte. Madrid: ITS Spain.

Datasets in Online Data Repository

- **Artuñedo, A.**, Godoy, J., Villagra, J. (2018). "Results of a comparison of traffic-free path planning primitives". IEEE Dataport. http://dx.doi.org/10.21227/H2W373

Acknowledgements

With this thesis I conclude an important academic and professional stage, in which I have been surrounded by many people who have contributed in one way or another to its development. For this reason, I take this opportunity to express my gratitude to each of them:

I would like to start from my two thesis directors, who since the beginning of our relationship, have been examples to follow and have made me grow both professionally and personally. On the one hand, Jorge Villagrá, who, since the end of my Master, has trusted and encouraged me to begin my journey in the research world. Thank you for your continuous support and time, for guiding me and always being clear and critical, and for motivating me in the most difficult moments. I consider myself lucky to be able to continue working and sharing good times with him. On the other hand, Rodolfo Haber, for those years of work in which he has transmitted his confidence and knowledge to me through wise advices, lessons and reflections, which have taught me to face problems from a broader perspective, thus facilitating the path.

Likewise, there are many people in the Centre for Automation and Robotics who have contributed to this thesis being carried out. Thus, I begin with those who have been closer during this time: Jorge Godoy, to whom I thank all I have learned about the previous work of AUTOPIA and his capacity and availability to solve so many problems that have been raised during the development of this thesis. Carlos González, for all the anecdotes and knowledge shared with me. To Raúl M. del Toro, who I had by my side in my early days in research, offering his support and having many good moments. Fernando Castaño and Gerardo Beruvides, who have made my welcome in Madrid more pleasant, sharing experiences and unforgettable trips. I do not forget the dining table, where Lola, Nacho, Fernando Seco, Antonio, Alberto, etc., have taken part in very interesting discussions, creating a good atmosphere (especially on Fridays). Finally, I do not want to ignore the rest of the CAR staff, who make our work easier.

My greatest thanks go to my family: my parents Antonio and Dolores, to whom I owe what I am today, and my brother Alvaro, who from German lands has always shown me his unconditional support. Thank you for your patience and understanding.

Finally, I would like to thank Mari Carmen. Her company in both good and bad moments has made this long journey easier. Thank you for motivating and supporting me from the very first moment.

Contents

Abbreviations

ABS Anti-lock Braking System
ACC Adaptive Cruise Control
ADAS Advanced Driver Assistance Systems
AEBS Advanced Emergency Braking System
API Application Programming Interface
AV Automated Vehicle
DBN Dynamic Bayesian Network
DGPS Differential Global Positioning System
ECU Electronic Control Unit
ESP Electronic Stability Program
GPS Global Positioning System
GPU Graphics Processing Unit
HMI Human Machine Interface
IMU Inertial Measurement Unit
IPC Inter-process Communication
IPM Inverse Perspective Mapping
ITS Intelligent Transportation Systems
LCA Lane Change Assist
LCM Lightweight Communication Marshalling
LDM Local Dynamic Map
LDW Lane Departure Warning
LiDAR Light Detection And Ranging
LKA Lane Keeping Assist
OS Operating System
OSM OpenStreetMap
PA Park Assist
PDC Park Distance Control
RTK Real Time Kinematic
TCS Traction Control System
UTM Universal Transverse Mercator

V2C	Vehicle-to (2)-Cloud
V2I	Vehicle-to (2)-Infrastructure
V2V	Vehicle-to (2)-Vehicle
V2X	Vehicle-to (2)-anything (X)
XML	Extensible Markup Language

Chapter 1
Introduction

1.1 Overview

Automated driving is considered to be one of the key technologies and major technological advances that influence and shape our future mobility and quality of life. So much so that most of the current activity in research and development in the field of Intelligent Transportation Systems (ITS) is focused on it. Events such as DARPA challenges [7, 12] or the Grand Cooperative Driving Challenge editions [8, 13] have raised the public interest in automated vehicles (AVs) worldwide.

The growing interest in ever higher levels of automation involves the development of algorithms capable of reacting to typical driving scenarios and making decisions to face increasingly complex driving situations in a safe and human-like manner. Although huge efforts have been carried out in last years to solve several technological challenges, still some issues must be addressed before AVs can be extensively deployed in urban environments [10]: (i) providing the automated driving system with comprehensive fault detection, identification and accommodation capabilities, (ii) ensuring sufficient cybersecurity protection, (iii) developing comprehensive environment perception capabilities, (iv) resolving questions of robot ethics sufficiently to enable the system software to make "life or death" decisions affecting the safety of all road users, and, finally (v) designing software-intensive system for a very high level of safety. In addition to that, the current advances in the AV technologies enable the arising of legal problems that need to be solved to encourage the societal acceptance of AVs.[1] Nevertheless, the potential benefits of AVs are more and more clearer, showing that AVs will improve the future mobility in different ways: more efficient and safer public transport and freight services, cleaner cities, accessibility and comfort.

---

[1]In this sense, a recent study reveals the growing interest of governments around the world on AVs [4].

A. Artuñedo, *Decision-making Strategies for Automated Driving in Urban Environments*, Springer Theses, https://doi.org/10.1007/978-3-030-45905-5_1

The number of vehicles on the roads is constantly growing, what affects the transport efficiency and increments traffic jams and carbon emissions. According to [9], the typically occupied area in a highway in the U.S. is around 5%, which means that the capacity of a road could be doubled or tripled if this area were increased by 10 or 15% factor, respectively. However, so as to achieve this, it is necessary to implement driver assistance systems to decrease the distance among cars in a safe way which also leads to a more efficient traffic. Moreover, lane width could be reduced by implementing automatic steering control systems with a higher degree of accuracy than human drivers have.

The introduction of automatic control systems in vehicles should greatly improve road safety taking into account that around 94% of traffic crashes are caused by human errors [11]. Nowadays, there are several solutions known as advanced driver assistance systems (ADAS) that are able to provide warning signals to the driver in some unexpected scenarios e.g. Lane Departure Warning (LDW) or Pedestrian Detection (PD), or even introducing longitudinal and/or lateral control actions in specific cases e.g. Automatic Cruise Control (ACC) or Lane Keeping Assist (LKA). The implementation of ADAS in commercial vehicles is more and more accepted by drivers, which have started to realize that the idea of automated vehicles is no longer a utopia and it is becoming a reality.

1.2 Motivation and Framework

From a technological point of view, automated driving involves the integration of a number of technologies related to perception, localization, decision-making and human-machine interaction (see Fig. 1.1). Among them, decision-making systems aim to provide the understanding of the vehicle environment, as well as generate a safe and efficient action plan in real-time [5]. Accordingly, tasks such as prediction of nearby traffic participants actions, motion planning, and obstacle avoidance must be carried out within the decision-making sub-systems.

Whilst considerable efforts have been attained in the perception and localization domains, digital representations of the world and planning in urban scenarios are still incomplete. As a result, understanding the spatio-temporal relationship between the subject vehicle and the relevant entities while being constrained by the road network is a very difficult challenge. Furthermore, urban motion planning is significantly affected since the knowledge about the environment is generally incomplete and the associated uncertainty is high. In addition, the predictions of future occupancy of nearby vehicles must effectively influence the motion plan calculated by the vehicle.

With the aim of reaching human-level abstract reasoning and reacting safely even in complex urban situations, autonomous driving requires methods to generalize unpredictable situations and reason in a timely manner. To that end, two elements still need further substantial investigation: world modelling and decision-making from uncertain information [14].

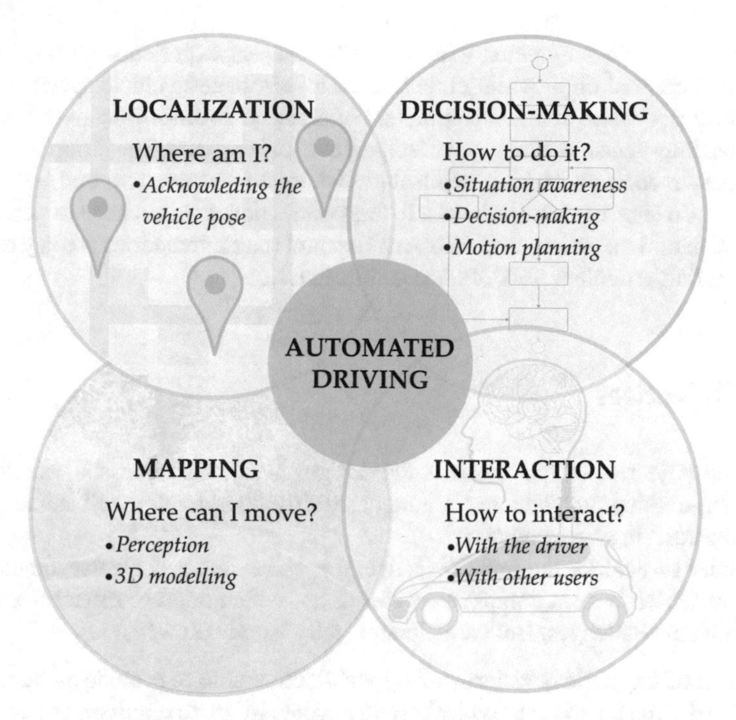

Fig. 1.1 Technologies involved in automated driving systems. Figure inspired by [2, p. 1279]

The objectives of this thesis are motivated by previous works on autonomous driving applications without detailed prior maps. In [6], the authors focus on rural environments where it is hard to built and maintain highly detailed maps. Despite of the fact that detailed maps in urban environments might be more feasible, they are still not cost-efficient and present technical problems as minor changes appear constantly, making them not completely reliable. Nevertheless, the information included in non-highly detailed maps remains more stable.

Bearing the above in mind, this thesis aims to contribute to the state-of-the-art in the decision-making and motion planning fields. In this regard, the thesis presents a modular architecture for automated driving in urban environments that provides both global and local planning capabilities that enable the vehicle to overcome successfully urban situations with certain complexity. Contrarily to some previous works on probabilistic decision-making [1, 3], often intractable in real-time, the setting presented in this work inherits the uncertainty from world modelling and produce feasible and comfortable trajectories which use cost-effective primitives evaluation and assuming simple patterns for targets motion prediction.

The present thesis takes place within the context of the AUTOPIA Program activities at the Centre for Automation and Robotics (CAR), a research centre with shared ownership between the Technical University of Madrid (UPM) and the Spanish National Research Council (CSIC). AUTOPIA is the pioneer research group in Spain focused on vehicle automation with more than 20 year experience on this field. Its research is centred on providing intelligence to automated vehicle systems in specific situations where communication and interaction abilities may permit to solve understanding-decision dilemmas of isolated self-driving cars. The group has a growing interest in decision-making architectures where driver intentions and skills can be adopted at different assistance levels. In this connection, the influence of perception, localization and mapping on decision-making and road interactions are key research questions that articulate AUTOPIA scientific activity.

1.3 Objectives

The main objective of this thesis is the design, implementation and validation of a decision-making architecture for automated driving specially well suited to cope with urban driving environments.

In order to achieve the main objective, this thesis focuses on five fundamental questions which in turn comprise several tasks, which address different scientific and technical challenges. Both are hierarchically listed below:

1. Design of a generic decision-making architecture able to provide an human-like and real-time response to typical driving events in urban environments:
    (a) Identification of essential architectural components to address the decision-making and planning in urban environments. The architecture must be able to provide solutions to typical driving problems faced in these environmental conditions.
    (b) Development of strategies with a high level of reactivity to provide fast responses to sudden critical situations that can arise from highly dynamic urban scenarios.
    (c) Elaboration of a functional human-machine interface to enable a bidirectional communication channel between the vehicle and its occupants.
2. Global planning capabilities and automatic road corridors generation from low-fidelity maps:
    (a) Development of a global planner that is able to provide high-level routes when requested by other modules of the architecture.
    (b) Obtaining a reliable available navigable space from the initial vehicle to the destination parting from high-level route and low-fidelity maps.
    (c) Adapting the navigable space by applying a vision-based approach with the objective of increasing its reliability.
3. Risk estimation and motion prediction of other traffic agents:
    (a) Explore interaction-aware risk estimation algorithms that can provide prioritization over the possible risks of other traffic agents perceived by the vehicle.

   (b) Development of motion prediction capability that estimates the future trajectory of the most hazardous agents, taking as input the risk prioritization.
4. Local planning capabilities that consider localization uncertainty and kinodynamic constrains of the vehicle:
   (a) Development of local planning capabilities to enable the vehicle to calculate human-like and human-aware trajectories in real time.
   (b) Considering the ego-vehicle constrains and other static or dynamic agents in the motion planning strategy in order to create feasible and safe trajectories.
5. Integration and validation of the whole architecture in a real environment:
   (a) Implementation and integration of all functionalities of the automated driving architecture.
   (b) Validation of all designed and implemented algorithms by using a real experimental platform in real-world scenarios.

1.4 Thesis Outline

A brief overview of the contents of the remaining chapters is presented below:

Chapter 2 presents a literature review of the different aspects of automated driving that are addressed in this thesis: map generation, risk estimation and motion planning.

Chapter 3 introduces the prior state of the automated driving architecture of AUTOPIA Program, remarking its benefits and limitations. Accordingly, the development of new functionalities and improvements on the existing architecture modules are then presented. The main contributions of this thesis concentrate on the decision-making capabilities of the architecture: on the one hand, global planning and mapping modules address deliberative features such as the computation of the route to reach a given destination. On the other hand, local planning modules deal with reactive decisions such as final trajectory generation and obstacle avoidance.

Chapter 4 details the contributions to the architecture regarding global planning capabilities. The global planner, which is in charge of providing high-level routes, is firstly introduced. From this global route, an automatic road corridor generation algorithm based on low-fidelity maps is proposed to compute the navigable space. Moreover, a vision-based road corridor adaptation algorithm is also proposed to increase the reliability of the navigable space. Finally, a method to consider the localization uncertainty using a probabilistic occupancy grid is introduced.

Chapter 5 deals with the estimation of future movements of relevant objects in the nearby environment of the ego-vehicle. To that end, an interaction-aware risk estimation algorithm uses information of traffic rules from maps and the state of the perceived objects, to select the most hazardous objects perceived by the vehicle. The output of the risk estimation algorithm is used to estimate the future trajectory of the most hazardous moving objects, which are then analysed to generate an output action if needed. The proposed risk estimation algorithm has been tested in simulation by using a set of different scenarios such as intersections, or roundabouts.

Chapter 6 addresses the optimal motion planning capabilities of the proposed architecture. Firstly, an in-depth analysis of state-of-the-art path planning approaches for automated driving is presented, where all considered approaches are evaluated and compared under a common framework. This extensive comparison leads to the selection of proper path and speed planning techniques to deal with the demanding requirements in terms of trajectory smoothness, space exploration and computing time. In this context, and with the goal of obtaining trajectories as human-like as possible, it is proposed a procedure to generate continuous curvature and jerk-minimum paths in real-time from automatically generated road corridors. The implemented trajectory generation algorithms are validated by a set performing real trials in urban-like scenarios by using a real instrumented vehicle.

Chapter 7 describes how the algorithms presented in Chaps. 4–6 are integrated in the architecture presented in Chap. 3. For that purpose, the detailed description of the experimental platform used for the implementation and the evaluation of the proposed architecture in real environments is firstly presented, focusing on both hardware and software aspects of the vehicle components. Eventually, two live demonstrations that were carried out at different high impact events in the autonomous driving and robotics fields are presented.

Chapter 8 presents the conclusions of the work carried out in this thesis together with specific contributions remarks. In addition, future research to continue the work is also proposed. Finally a summary of the dissemination activities related to this thesis is presented.

References

1. Brechtel S, Gindele T, Dillmann R (2014) Probabilistic decision-making under uncertainty for autonomous driving using continuous POMDPs. In: 17th international IEEE conference on intelligent transportation systems (ITSC), October 2014. IEEE, pp 392–399. ISBN 978-1-4799-6078-1. https://doi.org/10.1109/ITSC.2014.6957722, http://ieeexplore.ieee.org/document/6957722/
2. Eskandarian A (ed) (2012) Handbook of intelligent vehicles. Springer, London. ISBN 978-0-85729-084-7. https://doi.org/10.1007/978-0-85729-085-4, http://link.springer.com/10.1007/978-0-85729-085-4
3. Hubmann C, Schulz J, Becker M, Althoff D, Stiller C (2018) Automated driving in uncertain environments: planning with interaction and uncertain maneuver prediction. In: IEEE transactions on intelligent vehicles 3.1, March 2018, pp 5–17. ISSN 2379-8904. https://doi.org/10.1109/TIV.2017.2788208, http://ieeexplore.ieee.org/document/8248668/
4. KPMG International (2018) Autonomous vehicles readiness index. Technical report, p 60. https://assets.kpmg/content/dam/kpmg/xx/pdf/2019/02/2019-autonomous-vehicles-readiness-index.pdf
5. Liu S, Tang J, Zhang Z, Gaudiot JL (2017) Computer architectures for autonomous driving. In: Computer, vol 50, no 8, pp 18–25. ISSN 00189162. https://doi.org/10.1109/MC.2017.3001256, arXiv: 1702.01894

6. Ort T, Paull L, Rus D (2018) Autonomous vehicle navigation in rural environments without detailed prior maps. In: 2018 IEEE international conference on robotics and automation (ICRA), May 2018. IEEE, pp 2040–2047. ISBN 978-1-5386-3081-5. https://doi.org/10.1109/ICRA.2018.8460519, https://ieeexplore.ieee.org/document/8460519/
7. Patz BJ, Papelis Y, Pillat R, Stein G, Harper D (2008) A practical approach to robotic design for the DARPA urban challenge. J Field Robot 25(8):528–566. ISSN 15564959. https://doi.org/10.1002/rob.20251, arXiv: 10.1.1.91.5767, http://doi.wiley.com/10.1002/rob.20251
8. Ploeg J, Semsar-Kazerooni E, Morales Medina AI, de Jongh JFCM, van de Sluis J, Voronov A, Englund C, Bril RJ, Salunkhe H, Arrùe À, Ruano A, Garcìa- Sol L, van Nunen E, van de Wouw N (2018) Cooperative automated maneuvering at the 2016 grand cooperative driving challenge. In: IEEE transactions on intelligent transportation systems, vol 19, no 4, pp 1213–1226. ISSN 1524-9050. https://doi.org/10.1109/TITS.2017.2765669
9. Shladover S (2009) Cooperative (rather than autonomous) vehicle-highway automation systems. IEEE Intell Transp Syst Mag 1(1):10–19. ISSN 1939-1390. https://doi.org/10.1109/MITS.2009.932716, http://ieeexplore.ieee.org/document/5117654
10. Shladover SE (2018) Connected and automated vehicle systems: introduction and overview. J Intell Transp Syst 22(3):190–200. ISSN 1547-2450. https://doi.org/10.1080/15472450.2017.1336053
11. Singh S (2018) Critical reasons for crashes investigated in the national motor vehicle crash causation survey. (Traffic Safety Facts Crash Stats. Report No. DOT HS 812 506). National Highway Traffic Safety Administration, Washington, DC. https://crashstats.nhtsa.dot.gov/Api/Public/ViewPublication/812506
12. Thrun S, Montemerlo M, Dahlkamp H, Stavens D, Aron A, Diebel J, Fong P, Gale J, Halpenny M, Hoffmann G, Lau K, Oakley C, Palatucci M, Pratt V, Stang P, Strohband S, Dupont C, Jendrossek L-E, Koelen C, Markey C, Rummel C, van Niekerk J, Jensen E, Alessandrini P, Bradski G, Davies B, Ettinger S, Kaehler A, Nefian A, Mahoney P (2006) Stanley: the robot that won the DARPA grand challenge. J Field Robot 23(9):661–692. https://doi.org/10.1002/rob.20147, https://onlinelibrary.wiley.com/doi/abs/10.1002/rob.20147
13. van Nunen E, Kwakkernaat MRJAE, Ploeg J, Netten BD (2012) Cooperative competition for future mobility. In: IEEE transactions on intelligent transportation systems, September 2012, vol 13, no 3, pp 1018–1025. ISSN 1524-9050. https://doi.org/10.1109/TITS.2012.2200475
14. Watzenig D, Horn M (eds) (2017) Automated driving: safer and more efficient future driving. Springer International Publishing, Cham. ISBN 978-3-319-31893-6. https://doi.org/10.1007/978-3-319-31895-0, https://link.springer.com/10.1007/978-3-319-31895-0

Chapter 2
Literature Overview

2.1 Introduction

AVs can be defined as those in which at least some safety-critical aspects occur without direct driver input [74]. In other words, an AV is able to, at least partly, perform a driving task independently of a human driver. When AVs with different levels of automation can communicate among them and with the infrastructure/cloud, a very relevant socio-economic impact can be obtained, namely safety, congestion and pollution reduction, roads capacity increase, etc. By contrast, autonomous cars have theoretically the ability to operate independently and without the intervention of a human driver in a dynamic traffic environment, relying on the vehicle's own systems and without communicating with other vehicles or the infrastructure.

A wide variety of automated systems are continuously being developed by original equipment manufacturers (OEMs), Tier 1s, and new big players in the last years. Often, the level of automation of these systems is hard to differentiate since the manufactures usually use the term "autonomous" indistinctly. This has led different automotive associations to make efforts on the standardization of different driving automation levels. Among them, the most acknowledged framework is the one established by the Society of Automotive Engineers (SAE) International issued in 2014 in the international norm J3016 [64]. This norm brings order to different prior proposals of standardization from the US National Highway Traffic Safety Administration (NHTSA), the Germany Federal Highway Research Institute (BASt) and the SAE. It provides a common taxonomy for automated driving in order to simplify communication and facilitate collaboration within technical and policy domains.

Table 2.1 summarizes SAE levels of driving automation for on-road vehicles and their approximated correspondence to those developed by BASt and to those described by the NHTSA in its "*Preliminary Statement of Policy Concerning Automated Vehicles*" of May 30, 2013. Full definitions of these levels and detailed specifications of the concepts that appear in the table can be found in the Information Report J3016 [64]. As expressed by the expert committee responsible for this tax-

A. Artuñedo, *Decision-making Strategies for Automated Driving in Urban Environments*, Springer Theses, https://doi.org/10.1007/978-3-030-45905-5_2

Table 2.1 Summary of SAE levels [64]

Level	Name	Narrative definition	Execution of steering and acceleration/ deceleration	Monitoring of driving environment	Fallback performance of *dynamic driving task*	System capability (*driving modes*)	BASt level	NHTSA level
Human driver **monitors the driving environment**								
0	**No Automation**	the full-time performance by the *human driver* of all aspects of the *dynamic driving task*, even when enhanced by warning or intervention systems	Human driver	Human driver	Human driver	n/a	Driver only	0
1	**Driver Assistance**	the *driving mode*-specific execution by a driver assistance system of either steering or acceleration/deceleration using information about the driving environment and with the expectation that the *human driver* perform all remaining aspects of the *dynamic driving task*	Human driver and system	Human driver	Human driver	Some driving modes	Assisted	1
2	**Partial Automation**	the *driving mode*-specific execution by one or more driver assistance systems of both steering and acceleration/deceleration using information about the driving environment and with the expectation that the *human driver* perform all remaining aspects of the *dynamic driving task*	**System**	Human driver	Human driver	Some driving modes	Partially automated	2
Automated driving system **("system") monitors the driving environment**								
3	**Conditional Automation**	the *driving mode*-specific performance by an *automated driving system* of all aspects of the *dynamic driving task* with the expectation that the *human driver* will respond appropriately to a *request to intervene*	System	**System**	Human driver	Some driving modes	Highly automated	3
4	**High Automation**	the *driving mode*-specific performance by an *automated driving system* of all aspects of the *dynamic driving task*, even if a *human driver* does not respond appropriately to a *request to intervene*	System	System	**System**	Some driving modes	Fully automated	3/4
5	**Full Automation**	the full-time performance by an *automated driving system* of all aspects of the *dynamic driving task* under all roadway and environmental conditions that can be managed by a *human driver*	System	System	System	**All driving modes**	-	

Source [68]

onomy, "*the levels are descriptive rather than normative and technical rather than legal. They imply no particular order of market introduction. Elements indicate minimum rather than maximum system capabilities for each level. A particular vehicle may have multiple driving automation features such that it could operate at different levels depending upon the feature(s) that are engaged*".

According to Table 2.1 there are 6 levels that range from 0 (no automation) to 5 (full automation). In Level 0 systems, the human drives is responsible of the whole dynamic driving task. This level includes systems such as the Anti-lock Braking Systems (ABS), Traction Control System (TCS), Electronic Stability Program (ESP) and Advanced Emergency Braking Systems (AEBS); but also warning systems as Park Distance Control (PDC), Lane Change Assist (LCA) or Lane Departure Warning (LDW).

From the Level 1 onwards, capabilities of executing part of the dynamic driving task are introduced. Some examples of Level 1 systems are the Park Assist (PA), Lane Keeping Assist (LKA) or Adaptive Cruise Control (ACC).

To continue working on more complex systems, huge efforts are being made by the European automotive industry (about 5% of its total industry turnover [27]) in research and development of Level 2 systems. Level 2 implies partial automation so that both longitudinal and lateral control are provided but only in certain circumstances such as Automated Parking Assistance or Traffic Jam Assist, which can be considered as an extension of the ACC. Systems of this level need the human driver to be alert of the whole driving tasks as the system can need to be taken over in a very short notice by the human driver.

Level 3 is defined as *conditional automation*. This level introduces a clear different with respect to Level 2: now the system is in charge of the driving task monitoring.

Table 2.2 Differences between SAE Levels 4 and 5

		Some driving modes				
4	**High Automation**	Some geographic areas ⊕	Some roadway types ⊕	Some traffic conditions ⊕	Some weather conditions ⊕	Some events or incidents
		↓	↓	↓	↓	↓
5	**Full Automation**	**All** geographic areas ⊕	**All** roadway types ⊕	**All** traffic conditions ⊕	**All** weather conditions ⊕	**All** events or incidents*
		All driving modes				*that can be managed by a human driver

Figure inspired by [36]

This means that the system is expected to handle the driving as long as it is within its operational design domain, changing the role of the human driver to be a fall-back. For example, the Traffic Jam and Highway Chauffeurs are considered level 3 systems since the needed capability of performing both longitudinal and lateral control of the vehicle at moderate speeds (speeds below 60 km/h) in traffic jams at motorways.

The highly automated vehicles, which are meant to perform autonomous manoeuvres instead of supporting the driver, are framed in the levels 4 and 5. The main difference between these levels relies on the *unconditional* automated driving expected of Level 5 systems, which should be able to drive autonomously in the same scenarios that human drivers are, in contrast to the constrained operational design domain of Level 4 systems. These differences are illustratively summarized in Table 2.2. Nowadays, there is no Level 5 systems available. Nevertheless, some examples of Level 4 systems are the Parking Garage Pilot, which is able to perform all parking operations without the need of a human driver in a confined area, and the Highway Pilot in which the system performs all the driving tasks in a motorway without the need of a human driver.

In the industry sector it is remarkable the effort that some big companies have been making in last years. Google was the first big player which started its self-driving car project in 2009. Since then, Google has been a key member of an industry group pushing for automated driving standardization. At the end of 2016 Waymo became a stand-alone Google subsidiary in charge of the self-driving Car Project. In 2018, Waymo began the deployment of Level 4 vehicles in a pilot ride-sharing service in Chandler, Arizona. Waymos's system allows to drive at the speed limit it has stored on its maps and keep a safety distance from other vehicles in urban environments.

In 2015, Tesla introduced its "Autopilot" system which presents a combination of different ADAS that is able to accelerate, brake, and steer in specific situations. Its perception is based on RADARs and vision sensors. The system can be categorised as Level 2 since the driver's attention is required at all times. A similar approach is the "Super Cruise" proposed by General Motors. This system incorporates vision-based driver monitoring to detect whether the driver's eyes are on the road or not.

However, although the system allows hands-free driving, the driver alerts are issued at short notice and the driver must act quickly if necessary. This system presents a Level 2 automation.

Although there is a clear difference between Level 2 and 3 as stated above, it is not trivial to explain this difference in practice since Level 2 systems usually make use of some kind of driver monitoring too (e.g. General Motor's "Super Cruise"). Moreover, it is clear that Level 2 systems are not considered as self-driving systems while Level 4 do. Nevertheless, Level 3 is considered to provide self-driving capabilities in specific situations, giving the sufficient advance notices in cases in which the human driver is required. Thus, some automotive makers are moving away from Level 3 and focus on Level 2 or Level 4 systems while other such as Audi are proposing Level 3 systems. This is the case "Traffic Jam Pilot" developed by Audi. This system is intended to provide the human more time to take over the vehicle control than a Level 2 system would while performing self-driving in traffic jams.

The first tests carried out in public roads brought the first crashes and fatalities caused by self-driving cars: Three driver fatalities occurred between 2016 and 2018 in Tesla vehicles when the "Autopilot" functionality was activated; and an automated car operated by Uber struck and killed a woman on a street in Tempe, Arizona (March 2018). These accidents with fatal outcomes have questioned the readiness of automated driving systems to be massively deployed in public environments and have consequently led some car manufacturers (e.g. Uber or Toyota) to temporarily halt their testing of self-driving cars on public roads.

Although huge advances have been made in last years, still a number of technological issues and challenges have to be addressed to support the massive deployment of highly automated driving systems [36, 67]. They are listed below:

- **Decision and control algorithms**: These include decision, planning and control algorithms for a cooperative, safe, human compatible traffic automation. Moreover, "robot ethics" questions must be sufficiently resolved to take correct high level decisions in specific critical scenarios.
- **World modelling**: On the one hand, digital infrastructure for road automation includes static and dynamic digital representations of the physical world with which the automated vehicle will interact to operate. Issues to address include: sourcing, processing, quality control and information transmission. On the other hand, the development of comprehensive in-vehicle perception and prediction capabilities has to be addressed.
- **Secure V2X connectivity**: Connectivity is an important element of the automated vehicles especially secure V2X communication requiring low latency. V2X technologies encompass the use of wireless technologies to achieve real-time V2V and V2I communications. The convergence of sensor-based solutions (current ADAS) and V2X connectivity will promote automated driving. In addition, sufficient cyber-security is needed to ensure the safety of vehicle automated systems. When connected, the vehicle is susceptible to be cyber-attacked.
- **Human factors**: Human factors in automation relate to understanding the interaction(s) of humans with all aspects of an automated road transport system, both from

within a vehicle, when taking the role of a driver and also as a road user, when interacting with automated vehicles. Knowledge and theories from social, psychological and behavioural sciences are useful to understand how humans interact with such systems.

- **Evaluating road automation**: Automation of road vehicles has the potential to impact on lifestyles and society. Economic impacts too will be important and it will be necessary to gauge these impacts in a common cost-benefit framework with other transport investments when assessing public expenditure on supporting infrastructure or services.
- **Road-worthiness testing**: The evaluation whether a vehicle is legally allowed to drive on public roads or not takes great importance for the deployment of new automated driving functionalities. In this sense, AVs must include comprehensive fault detection systems and a very high level of safety has to be the basis of software-intensive system design.

Different research institutions worldwide are pushing to the above challenges needed to go towards a massive deployment of highly automated driving. An overview of the research work of some of them is provided below:

Carnegie Mellon University Since the 1980s, the Carnegie Mellon University have built computer-controlled vehicles for automated and assisted driving. They won the DARPA Urban Challenge (2007) robot race and placed third at the DARPA Robotics Challenge for disaster response robots (2015). Dr. Chris Urmson, one of the leaders of the winning team in the DARPA Urban Challenge, was part of the Google Car Project's research team from 2009 to 2016. More recent researches of this institution focus on smart infrastructure, focusing on machine vision application and affordable sensors.

California PATH PATH (*Partners for Advanced Transit and Highways*) has been involved in the research, development, and testing of connected and automated vehicles since 1986. The wide scope of California PATH work focuses on cooperative systems using V2X communication, cooperative intersection management, platooning, etc. Since 2015, California PATH is assisting the US administration by developing functional and technical definitions in automated vehicles regulations to help govern the testing, registration, and safe operation of autonomous vehicles.

Stanford University With the *Stanley* prototype they won the DARPA Grand Challenge (2005). The vehicle is a Volkswagen Touareg, where the native drive-by-wire control system was adapted to be run directly from an on-board computer without the use of actuators or servo motors. It used five roof mounted Lidars to build a 3D map of the environment, supplementing the position-sensing GPS system. *Junior* the sequel prototype of *Stanley*, obtained second place in the Urban Grand Challenge. The *Audi TTS* prototype managed to autonomously ascend Pikes Peak in 2010. More recently, they took the vehicle to Thunderhill Raceway Park, and let it go on track without anyone inside, hitting over 120 miles per hour. The goal of this prototype was to push autonomous driving to the vehicle's handling limits.

To that end, a high speed, consistent control signal is used in combination with numerous safety features capable of monitoring and stopping the vehicle.

Karlsruhe Institute of Technology This institute participated in the DARPA Urban Challenge (2005). Moreover, entered the VW Passat "AnnieWAY" into the Darpa Urban Challenge, and reaching the final stage. This vehicle also won the Grand Cooperative Driving Challenge (2011) which was the first international competition to implement highway platooning scenarios of cooperating vehicles connected with communication devices. Their research is focused on intelligent sensing techniques for autonomous vehicles and for automated visual inspection tasks. This includes scene reconstruction, object detection and tracking, accurate vehicle ego position estimation and automated map generation, probabilistic and formal representation, reasoning and learning techniques for scene interpretation.

Vislab Vislab, from the University of Parma, with which the International Autonomous Challenge was accomplished. The platforms are small and electric vehicles produced by Piaggio. The automated driving technology did not affect its performance since the sensors, the processing systems, and the actuation devices are all powered by solar energy, thus they do not drain anything from the original batteries. The vehicle managed to run almost 16.000 km on a 100-day trip, combining automated and manual mode in very challenging driving zones.

AUTOPIA Program Since 1996, AUTOPIA has a solid experience in providing intelligence to automated vehicle systems in specific situations where communication and interaction abilities may permit to solve understanding-decision dilemmas of isolated self-driving cars. The group has a growing interest in decision-making architectures where driver intentions and skills can be adopted at different assistance levels (from SAE L2 to L4). In this connection, the influence of perception, localization and mapping on decision-making and road interactions are key research questions that articulate AUTOPIA scientific activity [28].

The remainder of this chapter presents an overview of prior work on the different topics addressed in this thesis. Section 2.2 provides a general vision about the recent map-based systems. Furthermore, in Sect. 2.3 the focus is on the prediction of dynamic evolution of road agents and its influence on motion planners, taking into account information about the environment. Finally, Sect. 2.4 reviews the current state of motion planning in the automated driving scope.

2.2 Map Generation

An important number of map-related works for autonomous vehicles have been presented in the last decade. They range map-based localization and navigation [52, 61], acquisition, filtering and optimization techniques for high quality maps [10] or road modelling strategies [54]. However, not many combine smooth road geometry models and map-based advanced driving assistance systems (ADAS), whose state of the art is summarized below.

The most simple but commonly used representation for lanes in digital maps is based on poly-lines, composed of a sequence of segments, resulting in turn in concatenated polygons [80]. Among this family of mapping data structures, the RDDF was one of the first formats to describe route networks which was mainly used for unpaved desert tracks. It consisted of a simple list of longitudes, latitudes, and corridor widths that define the course boundaries, and a list of associated speed limits.

In RNDF the basic structure segment-lane-waypoint was included to provide basic information to driver-less vehicles. Multi-lane stretches of roads could be then modelled as road segments, defined by a set of waypoints, which in addition incorporates metadata to denote relevant elements, such as stop sign, intersection entry or exit, or checkpoints. Although the RNDF specification is fairly easy to understand and sufficient enough to map most road networks, it misses several features and reveals flaws when designing specific road characteristics [16]. Indeed, it is an efficient representation for real-time applications, but do not comply with real roads geometry, because they can exhibit discontinuities and often not describe curved structures in a suitable way. As a matter of fact, for a precise approximation of these structures, a relatively high number of line segments would be needed, which has two negative effects: On the one hand, the amount of data for storage and processing increases. And on the other hand, the association of points to their closest curve segment is hampered as the number of candidate segments is high.

OpenStreetMap (OSM) [62] proposes an open data infrastructure to which volunteers, companies or governmental organizations all over the world can contribute. Similarly to RNDF, OSM data is very good at representing topological information as well as positions, but has drawbacks in representing the geometry itself and good data accuracy is not guaranteed. Another open and freely available format is OpenDrive [21], which has enhanced features with respect to OSM, as it permits to mark roads being in a tunnel or on a bridge or even tilting and cross fall of roads. However, its deployment is still too limited when compared to the one of OSM, available worldwide with an impressive degree of coverage.

2.2.1 Smooth Road Geometry Models

The lanelet model also uses polylines as geometric representation but provides continuous tangents at the junctions of the segments thanks to non-Euclidean point-to-segment projection [6]. This approach provides an interesting real-time solution without discontinuity issues.

In RNDFGraph the course of the road is mimicked with spline interpolation. By using splines, only the support points for the spline need to be stored, and every point between the support points can be interpolated. The challenge, however, is to place the support points in such a way, so the resulting spline interpolation matches the course of the road. Support points are also the link to the graph. The RNDFGraph provides a road network with additional information for autonomous behavior such as continuous spline sampling, lane relationships and lane change information, to be used by subsequent path-planning and low-level controller modules.

Extended Maps [7] model the world in terms of interconnected clothoids, line and circle segments, following the Dubins paths structure [75]. Clothoids are used here to provide a smooth steering phase when passing the lane sections since their curvature changes linearly [77]. However, the offset curve of a clothoid is not a clothoid, and it is therefore not possible to easily build corridor using this primitive. Thus, it is reasonable to use alternative curves models, providing smooth continuous and differentiable curvilinear coordinates, such as approximative B-splines [42], Hermite interpolating splines [13, 33], NURBS [78], or arc splines [55].

The Akima interpolation presented in [16] is a continuously differentiable sub-spline interpolation whose resulting spline is less affected by outliers than cubic spline interpolation. According to [31] piecewise polynomials are more effective in terms of usability than clothoid or B-splines, because the tangent angle and the curvature of the road can be calculated by conducting simple arithmetic operations. However, the procedure proposed by the authors only works with positioning data, from which it determine the number of piecewise polynomials of the overall curve. In this connection, [53] propose an iterative method to automatically generate the lane geometry using fixed and variable control points, which can effectively ensure the accuracy with limited number of control points. Also in these two cases accurate samples of the centerline are required and therefore it is not suitable to enhance OSM maps.

2.2.2 Map-Based ADAS

ADASIS [22] is a standardization initiative for a data model to represent map data ahead of the vehicle (ADASIS Horizon) using a list of distinct potential future corridors, modeled as road segments. This standard does not store yet additional road features such as traffic lights or road signs, neither their position, as OSM-based systems do. However, they are not conceived as a map database and do not need to be queried to obtain the corridors, as happens with *lanelets* or similar approaches.

ADASIS allows also to obtain the centreline of the corridor as a smooth function, which is very useful for trajectory planning [3]. However, as this corridors are modelled as paths, it does not consider the possibility to characterize the lane margins independently of this centreline, which may really helpful in situations where the lane margins are asymmetrically distributed with respect to the lane centre.

2.3 Intention Prediction, Risk Estimation and Decision-Making

One of the key aspects to be solved in order to achieve fully autonomous driving is to find a risk indicator for any driving scene. An autonomous vehicle must recognize its environment and discern whether the behaviour of the vehicles around it is dangerous or not.

The focus of this review is on aspects related to the dynamic evolution of road users and its influence on motion planners, taking into account information about the environment (road maps, obstacles, etc.) and its associated uncertainty. Based on their capabilities to deal with motion planning on uncertain environments, two main families of approaches are explored: on the one hand, reward-based algorithms for motion planning and on the other hand, approaches based on risk inference techniques. The following subsections include further information about both approaches.

2.3.1 Reward-Based Algorithms for Motion Planning

These approaches are typically based on the application of Markov Decision Processes (MDPs) or Partially Observable MDP (POMDPs). Their ability in dealing with probabilistic uncertainty of the perceived environment make them an interesting topic in current research on automated driving. They provide a way to decide the optimal actions that the agents can take given a set of possible states, actions, reward functions, conditional transition probabilities and observations.

The dimensionality of most practical decision/planning problems is usually high, increasing the complexity of POMDPs which makes them computationally intractable in real-time. Due to the computational complexity of these approaches, they are usually used to compute the solution of the problem off-line for later application [4, 9]. Conversely, on-line methods need to make a trade-off between the solution quality and the complexity of the problem formulation, that depends on the state space size and the planning horizon [35]. Some examples found in the literature use POMDPs for motion planning in simple environments, where a robot can take long time to solve decision problems [24].

Although most of the (PO)MDP applications found in the literature deal with simple problem formulations that could be hardly translated to autonomous driving in a practical and useful way, some recent works propose real-time capable systems based on POMDPs. For example, the approach proposed in [35] explicitly includes the prediction uncertainty in terms of manoeuvre uncertainty and longitudinal uncertainty during the performed manoeuvre.

2.3.2 *Risk Based Motion Planning in Dynamic and Uncertain Environments*

Risk based motion planning approaches take into account the future trajectories of nearby vehicles in the motion planning. They comprise two main tasks: (i) intention prediction of moving obstacles and (ii) motion planning that considers the predicted trajectory of moving obstacles.

2.3.2.1 Trajectory Prediction and Risk Estimation of Moving Obstacles

Classic approaches for risk assessment in traffic scenes firstly predict the trajectory of the relevant vehicles using motion models. After that, it is checked if the trajectory of the ego-vehicle collides with the predicted ones [46]. Considering the uncertainties associated to the input data and the future possible events, the computation of all possibles trajectories becomes computationally expensive to be carried out in real-time. To overcome that, either the uncertainties are ignored or the independence between vehicles is assumed.

Recent approaches do not only rely on trajectory prediction to estimate the collision risk. As proposed in [48] the different approaches for risk estimation can be grouped in three main sets:

Physic-based methods These approaches use dynamic and kinematic models, control inputs, external conditions and vehicle physical properties to predict the future motion of the vehicle. The complexity of this approaches relies on the model used to predict the trajectory, increasing the complexity when dynamic models are employed. Kinematic models are commonly used making assumptions such as constant velocity (CV), constant acceleration (CA), constant turn rate and velocity (CTRV) or constant turn rate and acceleration (CTRA) in order to reduce the prediction complexity. The Kalman filter is one of the most used physic-based methods. The main drawback of these methods is their limited to very short-term motion prediction (typically less than a second).

Manoeuvre-based motion models These models represent vehicles as independent manoeuvring entities, assuming that the motion of a vehicle on the road network corresponds to a series of manoeuvres independently executed from the other vehicles. In manoeuvre-based motion models the trajectory prediction is based on the early recognition of the manoeuvres that drivers intend to perform. Different techniques are used in this methods: on the one hand they are based on prototype trajectories or manoeuvre intention estimation; on the other hand they are based on reachable sets:

- Support Vector Machines (SVM), Hidden Markov models (HMM), Gaussian processes (GP), Gaussian processes mixture (GPM): these methods comprise statistical models that need big amounts of statistical data for a initial training stage. After the training stage, the predictions are performed. Instead of predicting the real trajectory the vehicle will follow, these methods are usually used to obtain the probability that the vehicle will follow one the trajectories used to train the models.

- Reachable sets [1, 2]: The dynamics of moving obstacles are modelled off-line with Markov Chains, which are used on-line to predict reachable sets of tracked obstacles and provide an estimate of their potential future locations.

Interaction-aware motion models These methods take into account the inter-dependencies between vehicles' manoeuvres. The motion of a vehicle is assumed to be influenced by the motion of the other vehicles in the scene. Compared with the manoeuvre-based motion models, these methods leads to a better interpretation of their motion by taking into account the dependencies between the vehicles. Moreover, they contribute to a better understanding of the situation and a more reliable evaluation of the risk. Interaction-aware models allow longer-term predictions compared to physics-based motion models, and are more reliable than manoeuvre-based motion models since they account for the dependencies between the vehicles.

Among the three main sets of models presented above, interaction-aware motion models are the most complex and offer a better risk estimation with respect to manoeuvre-based and physics-based models since they take into account the intention of other agents. In fact, there are few interaction-aware motion models in the literature [47].

2.3.2.2 Approaches to Use Trajectory Prediction of Moving Obstacles in Motion Planning

As explained in [57], trajectory planning in dynamic environments is not as tractable as analogous problems in static environments. Some proposed approaches are based on numerically solving the motion planning problem by using variational methods directly in the time domain or by converting the trajectory planning problem to path planning problem by adding the time as a dimension of the configuration space [25]. Furthermore, a recent approach [69], proposes a method for identifying driving corridors in dynamic road scenarios by using a set representation of all reachable states. This approach makes possible to prove that certain high-level plans are infeasible. Some of the risk-based approaches for motion planning modify the classic rapidly exploring random tree algorithm (RRT) in different ways to take into account the prediction of other moving obstacles in the path planning process. In the case of the Fulgenci et al. approach [26], the likelihood of the obstacles' future trajectory and the probability of occupation are used to compute the risk of collision by using Gaussian processes.

2.4 Motion Planning

Among the decision-making tasks that an automated vehicle must carry out, motion planning is particularly relevant as it plays a key role in ensuring driving safety and comfort [40, 57] while producing safe, human-like and human-aware trajectories in a wide range of driving scenarios. The robotics community has been intensively working over the last 30 years in motion planning problems. Although many of the proposed algorithms are able to cope with a wide range of situations and contexts, they often demand computation-intensive algorithms, feasible for low speed motion patterns. However, for on-road autonomous driving, determinism is necessary at high sampling rates. In this context, optimality can be slightly sacrificed at the expense of safe human-adapted paths.

Two main drawbacks emerge from this approach: (i) these techniques often provide scenario-dependant solutions, which may cause wrong behaviours in general real driving on urban roads [29]; (ii) to guarantee reactivity, the trajectories need to be exhaustively sampled and evaluated in a high-dimensional space, which is computationally expensive. To cope with these limitations, some works (e.g. [30, 79]) propose a higher-level decision maker able to select the right cost set and sampling scale for different situations. Recent approaches are able to compute analytically both path and speed profile in real-time, taking into account kinematic and dynamic constraints of the vehicle [72].

The problem of finding an optimal path subject to holonomic constraints avoiding obstacles is known to be PSPACE-hard [11]. Significant research attention has been directed towards studying approximate methods or particular solutions of the general motion planning problem.

Since for most problems of interest in autonomous driving exact algorithms with practical computational complexity are unavailable [45], numerical methods are often used. These techniques generally do not find an exact answer to the problem, but attempt to find a satisfactory solution or a sequence of feasible solutions that converge to the optimal solution. The utility and performance of these approaches are typically quantified by the class of problems for which they are applicable as well as their guarantees for converging to an optimal solution. These approximate methods for path and trajectory planning can be divided in three main families [57]:

- Variational methods, that project the infinite-dimensional function space of trajectories to a finite-dimensional vector space. Direct methods approximate the solution to the optimal path with a set of parameters obtained with different types of non-linear continuous optimization techniques, often collocation-based integrators [84] or pseudoespectral approaches [17]. Indirect methods [8], in turn, solve the problem by finding solutions that satisfy the optimality conditions established by the Pontryagin's minimum principle [60].
- Graph-based search methods, that discretize the configuration space of the vehicle as a graph, where the vertices represent a finite collection of vehicle configurations and the edges represent transitions between vertices. The desired path is found by performing a search for a minimum-cost path in such a graph. There is a signifi-

cant number of strategies to construct that graph in the most efficient way, but they can be grouped in two main families: geometric methods, such as cell decomposition [12], visibility graphs [43] or Voronoi diagrams [71], and sampling-based methods [37, 41]. The latter deserves particular focus, as is particularly adapted to structured environments, where different steering functions (e.g. [58, 63]) or motion primitives (e.g. [23]) explore the reachability of the free configuration space. Once the graph is built, different strategies exist also to conduct the graph search in the most dependable way (e.g. Dijkstra [18], A* [32], D* [70], ...). In the case of automated driving, the road structure provides strong heuristics, where sampling-based planning methods are very often sufficient to produce a feasible solution [57]. An evolution of these methods, where spatio-temporal constraints are considered, propose to formulate the problem as a trajectory ranking and search problem, where multiple cost terms are combined to produce a specific behaviour.

- Incremental search methods sample the configuration space and incrementally build a reachability graph (often a tree) that maintains a discrete set of reachable configurations and feasible transitions between them. One of the most well-known and used techniques are the RRT [44] and their variants (e.g. [39]), always looking for the best trade-off between completeness and computational cost.

Since the applicability of variational methods is limited by their convergence to only local minima, graph-search methods try to overcome the problem by performing global search in a discretized version of the path space, generated by motion primitives. In some specific situations, this fixed graph discretization may lead to wrong or suboptimal solutions, in which case incremental search techniques may be useful, providing a feasible path to any motion planning problem instance, if one exists. In exchange, the required computation time to verify this completeness property may be unacceptable for a real-time system.

Given this context, a double stage planning approach, where a computationally-efficient reasonable traffic-free path is computed, seems the best trade-off for urban environments, where roads are usually well-structured and reasonably digitalized. In this connection, the most recent motion planning architectures [29, 30, 50] consider a two-step planning architecture that aims at limiting the search space to the region where the optimal solution is likely to exist, while keeping a high degree of reactivity. To that end, a first step performs the spatial space only considering the road geometry, resulting in a reference trajectory. Then, a traffic-based planning produces a path based on the traffic-free plan to account for other traffic and interfering objects. Finally, the final trajectory is generated for the most appropriate manoeuvre. Within this approach, a significant number of possible variations can significantly affect the resulting path. On this matter, the Subsect. 2.4.1 delves into the different path primitives used to compute the final path.

Heretofore, the reviewed motion planning approaches assume an ideal vehicle localization. However, different existing approaches take into account the localization uncertainty in the planning strategy. In this respect, Subsect. 2.4.2 gives an insight of these existing approaches that are proposed in the literature.

2.4.1 Path Primitives

In the literature many path generation patterns co-exist: Dubins' pioneering work [20] presented the first set of optimal primitive paths to go from point to point given initial and final orientations. Since them, some extensions have been introduced considering either linear/angular velocity bounds [5, 38] or proposing smoother curves (without curvature discontinuities) that permit to generate nearly time-optimal paths. Clothoids [75], arc-lines [34], spiral [51] and different variations of splines (e.g. [15, 56, 59, 76]) are some examples of them.

The criteria to choose the most adapted primitive are of course computational cost, safety and comfort, but also tunability and stability. Indeed, the resulting path has to be not only confined to the drivable space and needs to be compliant with fixed comfort-based acceleration and jerk bounds. Tunability also matters as designers may need to modify the planning strategy to avoid parameter over-fitting following the considered scenarios and/or different user preferences. In this connection, stability is also very important to avoid jerkiness in steering/braking when such context-based retuning is conducted.

In some cases, the curvature continuity cannot be guaranteed, and in many others the primitives are either difficult to parametrize or non-analytical, and therefore computationally expensive and/or unpredictable. Bézier curves have a closed-form expression and an intuitive way to choose their parameters. Although some previous work proposed combinations of symmetric curves [81] or smooth concatenations of cubic Bézier curves and segments [14], in this work uniform relaxed B-splines are used, concatenating continuous-curvature Bézier curves, so that more flexible solutions can be obtained.

Once the path primitive is chosen, the reference path planning becomes an optimization problem whose goal is to select the minimum number of waypoints needed to connect an origin and a destination, taking into consideration road geometry and comfort constraints. Multiple optimization criteria can be applied to that end [19, 73, 83]. However, the lack of an absolute trajectory quality indicator makes it hard to determine the most appropriate optimality criteria.

2.4.2 Considering Localization Uncertainty in Motion Planning

Localization plays an important role in autonomous driving since a certain level of accuracy in vehicle localization is indispensable for a safe navigation. Recent navigation systems rely on high-definition maps [49, 65, 66]. Assuming a high accuracy of the maps information, the localization with respect to the map plays an important role. Approaches as the proposed in [49] uses on-board sensors (such as LiDARs or cameras) to localize a moving vehicle relative to map, achieving a better localization accuracy than conventional GPS-IMU-odometry-based methods [49].

In some cases, the accuracy of localization systems can be low by design or even can drop depending on the environment e.g. GPS-based localization systems in cases in which there are reflections or satellites occlusions, or challenging weather conditions that could affect the positioning accuracy such as cloudy scenarios. In these situations, the localization uncertainty can be taken into consideration to increase the system reliability.

One interesting approach found in the literature to deal with the localization uncertainty when using maps is the one proposed in [82]. This paper focuses on encoding lane and traffic information in grids. This method requires the environment data to be represented in an occupancy grid, which provides a way to represent probabilistic information generated from different sensors measurements taking into account their noise and uncertainty. Then vehicle pose uncertainty is propagated along the occupancy grid, obtaining the occupancy probability of each cell of the grid. The authors use a prior map that stores detailed lane level information and then apply a localization uncertainty propagation algorithm over an occupancy grid. Occupancy grids are popular in autonomous navigation for encoding obstacle information into grid cells to provide real-time environmental models.

References

1. Althoff M, Stursberg O, Buss M (2009) Model-based probabilistic collision detection inautonomous driving. In: IEEE transactions on intelligent transportation systems, June 2009, vol 10, no 2, pp 299–310. ISSN 1524-9050. https://doi.org/10.1109/TITS.2009.2018966, http://ieeexplore.ieee.org/document/4895669/
2. Althoff M, Magdici S (2016) Set-based prediction of traffic participants on arbitrary road networks. In: IEEE transactions on intelligent vehicles, June 2016, vol 1, no 2, pp 187–202. ISSN 2379-8904. https://doi.org/10.1109/TIV.2016.2622920, http://ieeexplore.ieee.org/document/7725548/
3. Artuñedo A, Godoy J, Villagra J (2017) Smooth path planning for urban autonomous driving using OpenStreetMaps. In: 2017 IEEE intelligent vehicles symposium (IV), June 2017. IEEE, pp 837–842. ISBN 978-1-5090-4804-5. https://doi.org/10.1109/IVS.2017.7995820, http://ieeexplore.ieee.org/document/7995820/
4. Bai H, Hsu D, Lee WS (2014) Integrated perception and planning in the continuous space: a POMDP approach. In: Int J Robot Res 33(9):1288–1302. ISSN 0278-3649. https://doi.org/10.1177/0278364914528255, http://journals.sagepub.com/doi/10.1177/0278364914528255
5. Balkcom DJ, Mason MT (2002) Time optimal trajectories for bounded velocity differential drive vehicles. Int J Robot Res 21(3):199–217
6. Bender P, Ziegler J, Stiller C (2014) Lanelets: efficient map representation for autonomous driving. In: IEEE intelligent vehicles symposium (IV). IEEE, pp 420–425
7. Bétaille D, Toledo-Moreo R (2010) Creating enhanced maps for lane-level vehicle navigation. IEEE Trans Intell Transp Syst 11(4):786–798
8. Betts JT (1998) Survey of numerical methods for trajectory optimization. J Guid Control Dyn 21(2):193–207
9. Brechtel S, Gindele T, Dillmann R (2014) Probabilistic decision-making under uncertainty for autonomous driving using continuous POMDPs. In: 17th international IEEE conference on intelligent transportation systems (ITSC), October 2014. IEEE, pp 392–399. ISBN 978-1-4799-6078-1. https://doi.org/10.1109/ITSC.2014.6957722, http://ieeexplore.ieee.org/document/6957722/

10. Brummer S, Janda F, Maier G, Schindler A (2013) Evaluation of a mapping strategy based on smooth arc splines for different road types. In: 16th IEEE international conference on intelligent transportation systems (ITSC). IEEE, pp 160–165
11. Canny J, Reif J (1987) New lower bound techniques for robot motion planning problems. In: 28th annual symposium on foundations of computer science. IEEE, pp 49–60
12. Chazelle B (1985) Approximation and decomposition of shapes. Algorithmic Geom Asp Robot 1:145–185
13. Chen A, Ramanandan A, Farrell JA (2010) High-precision lane-level road map building for vehicle navigation. In: IEEE/ION position location and navigation symposium (PLANS). IEEE, pp 1035–1042
14. Choi J-W, Curry R, Elkaim G (2008) Path planning based on bézier curve for autonomous ground vehicles. In: Advances in electrical and electronics engineering-IAENG special edition of the world congress on engineering and computer science, WCECS'08. IEEE, pp 158–166
15. Connors J, Elkaim G (2007) Manipulating B-Spline based paths for obstacle avoidance in autonomous ground vehicles. In: Proceedings of the ION national technical meeting, vol 5. Citeseer
16. Czerwionka P, Wang M, Wiesel F (2011) Optimized route network graph as map reference for autonomous cars operating on German autobahn. In: 5th international conference on automation, robotics and applications (ICARA). IEEE, pp 78–83
17. Darby CL, Hager WW, Rao AV (2011) An hp-adaptive pseudospectral method for solving optimal control problems. Opt Control Appl Methods 32(4):476–502
18. Dijkstra EW (1959) A note on two problems in connexion with graphs. Numerische Mathematik 1(1):269–271
19. Dimitrakakis C (2006) Online statistical estimation for vehicle control. Technical report, IDIAP
20. Dubins LE (1957) On curves of minimal length with a constraint on average curvature, and with prescribed initial and terminal positions and tangents. Am J Math 79(3):497–516
21. Dupuis M et al (2010) Opendrive format specification. VIRES Simulations Technologie GmbH, Germany
22. Durekovic S, Smith N (2011) Architectures of map-supported ADAS. In: IEEE intelligent vehicles symposium (IV). IEEE, pp 207–211
23. Fleury S, Soueres P, Laumond J-P, Chatila R (1995) Primitives for smoothing mobile robot trajectories. IEEE Trans Robot Autom 11(3):441–448
24. Foka A, Trahanias P (2007) Real-time hierarchical POMDPs for autonomous robot navigation. In: Robotics and autonomous systems, July 2007, vol 55, no 7, pp 561–571. ISSN 0921-8890. https://doi.org/10.1016/j.robot.2007.01.004, https://linkinghub.elsevier.com/retrieve/pii/S0921889007000279
25. Fraichard T (1999) Trajectory planning in a dynamic workspace: a 'state-time space' approach. In: Advanced robotics, January 1999, vol 13, no 1, pp 75–94. ISSN 01691864. https://doi.org/10.1163/156855399X01017. http://openurl.ingenta.com/content/xref?genre=article5C&
26. Fulgenzi C, Spalanzani A, Laugier C, Tay C (2010) Risk based motion planning and navigation in uncertain dynamic environment. Research report 14, October 2010, pp 1–14. ISSN 03029743. https://doi.org/10.1007/978-3-540-71496-5_4, https://hal.inria.fr/inria-00526601/
27. Gleave D, Frisoni R, Dall'Oglio A, Nelson C, Long J, Vollath C, Ranghetti D, McMinimy S (2016) Self-piloted cars: the future of road transport? European Union. ISBN 978-92-823-9056-6. https://doi.org/10.2861/66390, http://www.europarl.europa.eu/thinktank/en/document.html?reference=IPOL
28. Godoy J, Artuñedo A, Haber R, González C (2015) Conducción autónoma y cooperative—El programa Autopia en España. In: XV Congreso Español sobre Sistemas Inteligentes de Transporte, April 2015. ITS Spain, Madrid
29. Gu T, Dolan JM, Lee J-W (2016) On-road trajectory planning for general autonomous driving with enhanced tunability. In: Advances in intelligent systems and computing, pp 247–261. ISBN 9783319083377. https://doi.org/10.1007/978-3-319-08338-4_19. http://link.springer.com/10.1007/978-3-319-08338-4

30. Gu T, Snider J, Dolan JM, Lee J-W (2013) Focused trajectory planning for autonomous on-road driving. In: 2013 IEEE intelligent vehicles symposium (IV), June 2013. IEEE, pp 547–552. ISBN 978-1-4673-2755-8. https://doi.org/10.1109/IVS.2013.6629524, http://ieeexplore.ieee.org/document/6629524/
31. Gwon G-P, Hur W-S, Kim S-W, Seo S-W (2017) Generation of a precise and efficient lane-level road map for intelligent vehicle systems. IEEE Trans Vehic Technol 66(6):4517–4533
32. Hart PE, Nilsson NJ, Raphael B (1968) A formal basis for the heuristic determination of minimum cost paths. IEEE Trans Syst Sci Cybern 4(2):100–107
33. Héry E, Masi S, Xu P, Bonnifait P (2017) Map-based curvilinear coordinates for autonomous vehicles. In: 20th IEEE intelligent transportation systems conference (ITSC)
34. Horst J, Barbera A (2006) Trajectory generation for an on-road autonomous vehicle. In: Gerhart GR, Shoemaker CM, Gage DW (eds) Unmanned systems technology VIII. May 2006, p 62302J. https://doi.org/10.1117/12.663643, http://proceedings.spiedigitallibrary.org/proceeding.aspx?doi=10.1117/12.663643
35. Hubmann C, Schulz J, Becker M, Althoff D, Stiller C (2018) Automated driving in uncertain environments: planning with interaction and uncertain maneuver prediction. In: IEEE transactions on intelligent vehicles, March 2018, vol 3, no 1, pp 5–17. ISSN 2379-8904. https://doi.org/10.1109/TIV.2017.2788208, http://ieeexplore.ieee.org/document/8248668/
36. International Transport Forum (2015) Automated and autonomous driving: regulation under uncertainty. International Transport Forum Policy Papers 7 (2015). https://doi.org/10.1787/5jlwvzdfk640-en, https://www.oecd-ilibrary.org/content/paper/5jlwvzdfk640-en
37. Janson L, Schmerling E, Clark A, Pavone M (2015) Fast marching tree: a fast marching sampling-based method for optimal motion planning in many dimensions. Int J Robot Res 34(7):883–921
38. Jiang K, Seneviratne LD, Earles S (1997) Time-optimal smooth-path motion planning for a mobile robot with kinematic constraints. Robotica 15(05):547–553
39. Karaman S, Frazzoli E (2010) Optimal kinodynamic motion planning using incremental sampling-based methods. In: 2010 49th IEEE conference on decision and control (CDC). IEEE, pp 7681–7687
40. Katrakazas C, Quddus M, Chen WH, Deka L (2015) Real-time motion planning methods for autonomous on-road driving: state-of-the-art and future research directions. In: Transportation research part C: emerging technologies, November 2015, vol 60, pp 416–442. ISSN 0968090X. https://doi.org/10.1016/j.trc.2015.09.011, https://linkinghub.elsevier.com/retrieve/pii/S0968090X15003447
41. Kavraki LE, Svestka P, Latombe J-C, Overmars MH (1996) Probabilistic roadmaps for path planning in high-dimensional configuration spaces. IEEE Trans RobotAutom 12(4):566–580
42. Koutaki G, Uchimura K, Hu Z (2006) Network active shape model for updating road map from aerial images. In: IEEE intelligent Vehicles Symposium (IV). IEEE, pp 325–330
43. Latombe J-C (2012) Robot motion planning, vol 124. Springer Science & Business Media, New York
44. LaValle SM (1998) Rapidly-exploring random trees: a new tool for path planning. Technical Report 98-11, Department of CS, Iowa State University. http://msl.cs.illinois.edu/~lavalle/papers/Lav98c.pdf
45. Lazard S, Reif J, Wang H (1998) The complexity of the two dimensional curvatureconstrained shortest-path problem. In: Proceedings of the third international workshop on the algorithmic foundations of robotics, Houston, TX, USA. Citeseer, pp 49–57
46. Lefàvre S (2012) Risk estimation at road intersections for connected vehicle safety applications. Theses, Université de Grenoble, October 2012. https://tel.archives-ouvertes.fr/tel-00858906
47. Lefevre S, Laugier C, Ibanez-Guzman J (2012) Evaluating risk at road intersections by detecting conflicting intentions. In: 2012 IEEE/RSJ international conference on intelligent robots and systems, October 2012. IEEE, pp 4841–4846. ISBN 978-1-4673-1736-8. https://doi.org/10.1109/IROS.2012.6385491, http://ieeexplore.ieee.org/document/6385491/
48. Lefàvre S, Vasquez D, Laugier C (2014) A survey on motion prediction and risk assessment for intelligent vehicles. ROBOMECH J 1(1):1. ISSN 2197-4225. https://doi.org/10.1186/s40648-014-0001-z,abs/1607.04788, http://www.robomechjournal.com/content/1/1/120, http://www.ncbi.nlm.nih.gov/pubmed/1569940nih.gov/articlerender.fcgi?artid=PMC364376

49. Levinson J, Montemerlo M, Thrun S (2007) Map-based precision vehicle localization in urban environments. In: Robotics: science and systems III, June 2007. ISBN 9780262524841. https://doi.org/10.15607/RSS.2007.III.016, http://www.roboticsproceedings.org/rss03/p16.pdf
50. Li X, Sun Z, Cao D, He Z, Zhu Q (2016) Real-time trajectory planning for autonomous urban driving: framework, algorithms, and verifications. IEEE/ASME Trans Mechatr 21(2):740–753
51. Liang T-C, Liu J-S, Hung G-T, Chang Y-Z (2005) Practical and flexible path planning for car-like mobile robot using maximal-curvature cubic spiral. Robot Auton Sys 52(4):312–335
52. Liu C, Jiang K, Yang D, Xiao Z (2017) Design of a multi-layer lane-level map for vehicle route planning. In: MATEC web of conferences, vol 124. EDP Sciences, p 03001
53. Liu J, Cai B, Wang Y, Wang J (2013) Generating enhanced intersection maps for lane level vehicle positioning based applications. Procedia-SocBehav Sci 96:2395–2403
54. Liu L, Wu T, Fang Y, Hu T, Song J (2015) A smart map representation for autonomous vehicle navigation. In: 12th international conference on fuzzy systems and knowledge discovery (FSKD). IEEE, pp 2308–2313
55. Meek D, Walton D (2004) An arc spline approximation to a clothoid. J Comput Appl Math 170(1):59–77
56. Nelson W (1989) Continuous-curvature paths for autonomous vehicles. In: Proceedings of the IEEE international conference on robotics and automation. IEEE, pp 1260–1264
57. Paden B, Cáp M, Yong SZ, Yershov D, Frazzoli E (2016) A survey of motion planning and control techniques for self-driving urban vehicles. In: IEEE transactions on intelligent vehicles, March 2016, vol 1, no 1, pp 33–55. ISSN 2379-8904. https://doi.org/10.1109/TIV.2016.2578706
58. Petti S, Fraichard T (2005) Safe motion planning in dynamic environments. In: 2005 IEEE/RSJ international conference on intelligent robots and systems (IROS 2005). IEEE, pp 2210–2215
59. Piazzi A, Bianco CGL, Romano M (2007) h3 Splines for the smooth path generation of wheeled mobile robots. IEEE Trans Robot 23(5):1089–1095
60. Pontryagin LS (1987) Mathematical theory of optimal processes. CRC Press, Boca Raton
61. Rabe J, Meinke M, Necker M, Stiller C (2016) Lane-level map-matching based on optimization. In: 19th IEEE international conference on intelligent transportation systems (ITSC). IEEE, pp 1155–1160
62. Ramm F, Topf J, Chilton S (2011) OpenStreetMap: using and enhancing the free map of the world. UIT Cambridge, Cambridge
63. Reeds J, Shepp L (1990) Optimal paths for a car that goes both forwards and backwards. Pac J Math 145(2):367–393
64. SAE International (2018) Taxonomy and definitions for terms related to driving automation systems for on-road motor vehicles. Technical report. https://doi.org/10.4271/J3016_201806
65. Schindler A (2013) Vehicle self-localization with high-precision digital maps. In: 2013 IEEE intelligent vehicles symposium workshops (IV Workshops), June 2013. IEEE, pp 134–139. ISBN 978-1-4799-0795-3. https://doi.org/10.1109/IVWorkshops.2013.6615239, http://ieeexplore.ieee.org/document/6615239/
66. Seif HG, Hu X (2016) Autonomous driving in the iCity-HD maps as a key challenge of the automotive industry. Engineering. ISSN 20958099. https://doi.org/10.1016/J.ENG.2016.02.010
67. Shladover SE (2018) Connected and automated vehicle systems: introduction and overview. J Intell Transp Syst 22(3):190–200. ISSN 1547-2450. https://doi.org/10.1080/15472450.2017.1336053
68. Smith BW (2013) SAE levels of driving automation, December 2013. http://cyberlaw.stanford.edu/blog/2013/12/sae-levels-driving-automation
69. Sontges S, Althoff M (2017) Computing possible driving corridors for automated vehicles. In: 2017 IEEE intelligent vehicles symposium (IV), June 2017. IEEE, pp 160–166. ISBN 978-1-5090-4804-5. https://doi.org/10.1109/IVS.2017.7995714, http://ieeexplore.ieee.org/document/7995714/
70. Stentz A (1994) Optimal and efficient path planning for partially-known environments. In: Proceedings of IEEE international conference on robotics and automation. IEEE, pp 3310–3317

71. Takahashi O, Schilling RJ (1989) Motion planning in a plane using generalized Voronoi diagrams. IEEE Trans Robot Autom 5(2):143–150
72. Talamino JP, Sanfeliu A (2019) Anticipatory kinodynamic motion planner for computing the best path and velocity trajectory in autonomous driving. In: Robotics and autonomous systems, vol 114, pp 93–105. ISSN 0921-8890. https://doi.org/10.1016/j.robot.2018.11.022. http://www.sciencedirect.com/science/article/pii/S0921889018301957
73. Tran DQ, Diehl M (2009) An application of sequential convex programming to time optimal trajectory planning for a car motion. In: Proceedings of the 48th IEEE conference on decision and control, 2009 held jointly with the 2009 28th Chinese control conference, CDC/CCC 2009. IEEE, pp 4366–4371
74. Villagra J, Acosta L, Artuñedo A, Blanco R, Clavijo M, Fernández C, Godoy J, Haber R, Jiménez F, Martínez C, Naranjo JE, Navarro PJ, Paúl A, Sánchez F (2018) Automated driving. In: Intelligent vehicles. Elsevier, pp 275–342. ISBN 9780128128008. https://doi.org/10.1016/B978-0-12-812800-8.00008-4, http://linkinghub.elsevier.com/retrieve/pii/B9780128128008000084
75. Villagra J, Milanés V, Pérez J, Godoy J (2012) Smooth path and speed planning for an automated public transport vehicle. In: Robotics and autonomous systems, February 2012, vol 60, no 2, pp 252–265. ISSN 09218890. https://doi.org/10.1016/j.robot.2011.11.001, https://linkinghub.elsevier.com/retrieve/pii/S092188901100203X
76. Villagra J, Mounier H (2005) Obstacle-avoiding path planning for high velocity wheeled mobile robots. In: IFAC proceedings volumes (IFAC-PapersOnline), vol 38 no 1. Elsevier, pp 49–54. ISBN 008045108X
77. Walton DJ, Meek DS (2005) A controlled clothoid spline. Comput Graph 29(3):353–363
78. Wang L, Miura KT, Nakamae E, Yamamoto T, Wang TJ (2001) An approximation approach of the clothoid curve defined in the interval [0, p/2] and its offset by freeform curves. Compu-Aided Des 33(14):1049–1058
79. Wei J, Snider JM, Gu T, Dolan JM, Litkouhi B (2014) A behavioral planning framework for autonomous driving. In: 2014 IEEE intelligent vehicles symposium proceedings, June 2014. IEEE, pp 458–464. ISBN 978-1-4799-3638-0. https://doi.org/10.1109/IVS.2014.6856582, http://ieeexplore.ieee.org/document/6856582/
80. Xu Y, Sasse V, Harms K (1996) The European digital road map mutimap and its applications. Int Arch Photogram Remote Sens 31:982–987
81. Yang K, Sukkarieh S (2010) An analytical continuous-curvature path-smoothing algorithm. IEEE Trans Robot 26(3):561–568
82. Yu C, Cherfaoui V, Bonnifait P (2016) Semantic evidential lane grids with prior maps for autonomous navigation. In: IEEE conference on intelligent transportation systems, proceedings, ITSC, pp 1875–1881. ISBN 9781509018895. https://doi.org/10.1109/ITSC.2016.7795860
83. Ziegler J, Bender P, Dang T, Stiller C (2014) Trajectory planning for Bertha—a local, continuous method. In: 2014 IEEE intelligent vehicles symposium proceedings. IEEE, pp 450–457
84. Ziegler J, Bender P, Schreiber M, Lategahn H, Strauss T, Stiller C, Dang T, Franke U, Appenrodt N, Keller CG et al (2014) Making Bertha drive—an autonomous journey on a historic route. IEEE Intell Transp Syst Mag 6(2):8–20

Chapter 3
Decision-Making Architecture

3.1 Introduction

The architecture design to provide automated driving capabilities plays a key role since it must include all needed components that allow the vehicle to have the expected behaviour in the driving scenarios for which it has been designed. These driving scenarios are more and more demanding as they include a growing number of inter-connected and heterogeneous agents and the associated caseload may become intractable. Thus, the design of the architecture must be made bearing in mind the perfect integration of its elements.

In this sense, this chapter provides the description of the prior automated driving architecture of AUTOPIA and the main contributions to it that are proposed in this thesis. Firstly, the prior developments of the architecture are introduced in Sect. 3.2. Then, the new components proposed in this work are presented in Sect. 3.3, emphasising their motivations and integration within the architecture.

3.2 Prior State of the Architecture

One of the most important considerations taken into account in the design of the prior architecture was the system modularity. In this sense, all functionalities were devised as different modules that interact with others. Regarding the implementation, both hardware and software components are implemented as different modules. The advantage of this fragmentation is twofold: (i) the development of different modules can be done in parallel without interfering with each other, and in case of a system failure the problematic modules can be quickly identified. (ii) If each software module is defined as an independent program, a failure in one module does not necessarily affect the whole system.

A. Artuñedo, *Decision-making Strategies for Automated Driving in Urban Environments*, Springer Theses, https://doi.org/10.1007/978-3-030-45905-5_3

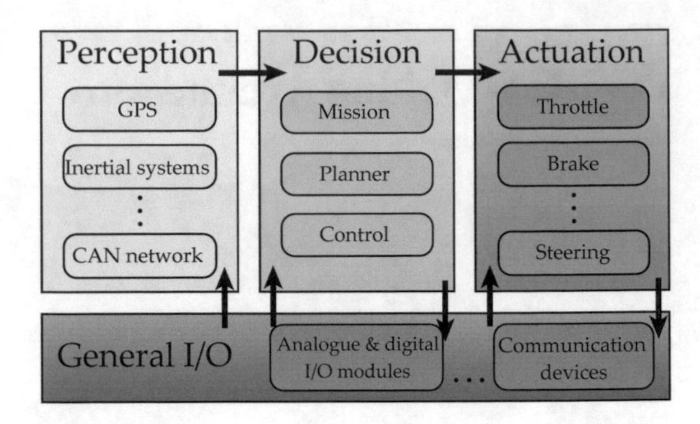

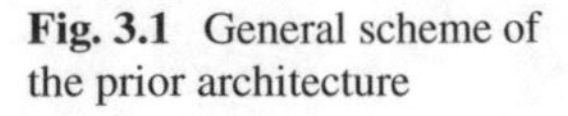
Fig. 3.1 General scheme of the prior architecture

The prior AUTOPIA architecture was divided in three main stages: **Perception**, **Decision** and **Actuation**; and an auxiliary one shared by the latter three: **General I/O**. As can be seen in Fig. 3.1, the **Perception** stage was in charge of providing information about vehicle state and its surroundings, i.e. other vehicles or infrastructure. The **Decision** stage includes goals, motion planning, vehicle state management and determination of the control actions. The **Actuation** stage was composed of the systems executing the control actions over the throttle, brake and steering wheel of the vehicle. Finally, the **General I/O** includes all the low-level devices, such as communications and digital/analogue I/O modules, that can be used by any other module of the main layers. This removes the direct dependency between a hardware module and a unique software module, which allows several software modules can use this hardware resource.

Prior developments [2] have provided the **Perception**, **Actuation**, and **General I/O**, stages with full vehicle control functionalities and simple planning capabilities. Which are detailed in Sect. 7.2. Regarding the **Decision** stage, core of this work, the prior architecture included three different modules: a mission module, planner module and control module, which are described in the following subsections.

3.2.1 Mission and Planner Modules

The main task of a self-driving car can be summarized as "going from point A to point B". Besides, it can be more complex as for a self-driving bus that must follow a pre-defined route, making several stops and waiting while the passengers go up and down. In any of these two cases, the main goal can be defined as a combination of several smaller sub-goals. The set of these sub-goals is what it is called the mission of the car.

In the prior architecture, the mission module represents the first step in the decision stage. Inside this module, a list of generic goals, defined manually by the user either before running the program or even in running time, describes the major goal for the car.

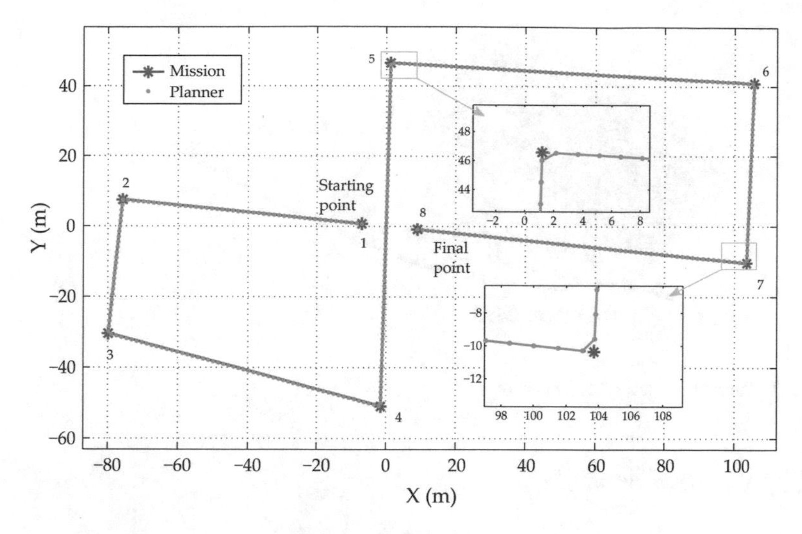

Fig. 3.2 Planner points generated from the mission data

Once a mission is given, the next step is to generate an adequate trajectory for the car to perform the main task. As it was mentioned above, the mission module only has the information related to the main goal of the car, meaning that the mission list can be composed of only two points: start and goal, or of a long list of points describing the movement of a leader vehicle (in case of ACC manoeuvres). To analyse the mission and generate the final trajectory, a planner module is included.

The trajectory was composed of the reference points information such as UTM coordinates, the reference speed, the cumulative distance from the beginning of the trajectory, and other related data. The planner might need information about the road network in order to know the best path for going from one point to another of the map, depending on the trajectory task requirements. Once the route to go from one mission point to another is set, the planner module computes equidistant points along the route. These points conforms the final trajectory the vehicle will follow.

An example of a mission and planned path with the prior architecture is shown in Fig. 3.2. In this case, the distance between the planned points is equal to 1.5 m (approximately a half of the vehicle length). In order to perform lateral control, the initial reference is determined by taking the closest two points to the actual position of the car and using the segment they define as trajectory reference. As the car moves, the following reference points are loaded and the older ones dismissed. Despite being able to describe a straight line using only two points (start and end), the long segments are divided into smaller ones of equal length. This allows the planner to modify, if necessary, the trajectory ahead of the current position even when the vehicle is already on a straight line, avoiding to change a reference in use.

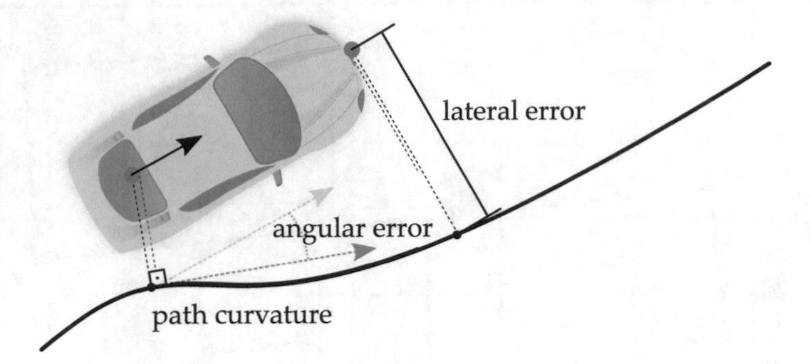

Fig. 3.3 Variables used lateral control

3.2.2 *Control Module*

The last step of the decision stage is the computation of the control variables, control mode and control actions. As can be found in previous works [3, 4, 7, 9], the approach followed by the AUTOPIA Program was not the development of a global controller able to perform all the possible scenarios, but the individual analysis the common ones, e.g. overtaking, merging, ACC, emergency braking, etc., developing a specific controller for each one. The choice of these controllers was called the control mode.

In order to keep the software execution time as low as possible, the computation of the control variables was divided in two phases. During the first phase, the program calculated only the variables required for the selection of the control mode as the speed, angular and lateral errors. Once the control mode was selected, the program estimated the mode-specific variables as the time gaps, relative speeds, priorities and so on. The list of variables calculated in each phase was user-customizable should new modes be added.

The generation of the control actions for driving the vehicle has been widely covered by the AUTOPIA team in previous works [5, 6, 8]. For lateral control, the simplest approach consists in the generation of control actions from both lateral and angular errors measured from the reference path to vehicle position and orientation respectively (Fig. 3.3). Regarding the longitudinal control, a speed controller is able to generate the low-level commands for throttle and brake, following the speed references for each area.

3.2.3 *Prior Architecture Remarks*

As described above, prior developments on the architecture introduced a modular architecture for vehicle automation that presents significant advantages:

1. The software is not centralized in a master process, but it is composed of a set of smaller processes running simultaneously. This exploits in a better way the computation resources and minimizes the risk of deadlocks and processing delays [10].
2. Moreover, by dividing the program in several software modules, it is possible to minimize the impact of a failure in one of the modules over the global performance of the system. For example, if a supervisor program is implemented, only the crashed process is restarted as soon as the failure is detected, without being necessary to restart all the software.
3. Although the software fragmentation in several processes required the usage of an inter-process communication (IPC) method, the use of the LCM library (which is detailed in Sect. 7.2.2.1) increases the capability of the developers for debugging and detecting system failures. Furthermore, thanks to the time stamps, all the system data can be logged and replayed off-line as it was sent through the network.
4. When implementing the architecture in other vehicles, modifications could be only needed in some of the sensor/actuator modules, reducing the implementation time.

Although the previous works on the architecture present a remarkable design and hardware-software integration, the decision-making architecture capabilities can be extended by pushing some limitations that restrict the application field to specific driving cases and by adding new features. The following list present the identified limitations categorised in the main technology groups evoked in Fig. 1.1:

1. **Localization**

   (a) The localization of the vehicle relies mainly on raw RTK-GPS data. Despite the positioning accuracy is high, the orientation of the vehicle is not reliable at low speeds.

2. **Mapping**

   (a) The lack of exteroceptive sensors such as camera or LiDAR constrains the vehicle perception of the nearby environment.
   (b) The architecture does not use information from any map sources. This limits the knowledge of the environment and lead the developed systems to rely on external information received through V2X communication and predefined off-line computed paths.

3. **Decision-making**

   (a) The absence of environmental information restricts the decision-making and planning capabilities of the architecture, making the vehicle unable to react to events occurred in typical driving scenarios such as pedestrian crossings, obstacles that must be avoided, etc.
   (b) By not using maps, the vehicle is unable to compute by itself a high-level route to reach the required destination.
   (c) The simple motion planning algorithm outputs large sections of straight segments whose junctions implies abrupt changes in lateral control error.

This limits the performance of lateral controllers of the vehicle since they must be designed to handle these non-smooth inputs. In consequence, high path tracking error is obtained.

(d) The simple motion planning does not include an speed planner. Just like in the case of lateral controllers, the longitudinal ones are also designed to handle non-smooth speed reference inputs, leading a low performance of the longitudinal controller.

4. **Interaction**
   (a) The lack of an HMI element makes it impossible for end users to interact in an easy and friendly way with the vehicle. The existing interaction between the vehicle computer and the human is done at low level computer programs that need the presence of the developers.
   (b) The prior architecture does not include a way to visualize the state and decisions of the vehicle during its operation.

3.3 Contributions to the Architecture

On the basis of previous works on the AUTOPIA architecture and in view of the limitations described above, the present thesis addresses different improvements and extensions mainly in its decision-making stage, increasing the general navigation capability of the system. Moreover, several modifications and adjustments have been carried out in some of the existing modules in order to be seamlessly integrates with the new ones.

The proposed contributions to the architecture consist of a combination of new modules with different functionalities that provide the whole system with automated driving capabilities [1]. These components are depicted in the general functional diagram of Fig. 3.4, where dashed lines represent event-driven actions and continuous lines correspond to continuously applied actions. Henceforth, this figure is referenced to provide the general view of the connections among the different high level modules of the architecture.

On the one hand, global planning and mapping modules address deliberative features such as the computation of the route to reach a given destination. On the other hand, local planning modules deal with reactive decisions such as final trajectory generation and obstacle avoidance.

The following list summarizes both (i) the contributions by gathering the new functional components added to the architecture, which are introduced in the subsections below, and (ii) the improvements on other already implemented modules:

- Global planning functionalities:
  - Road corridor generation from OSM
  - Vision-based road corridor adaptation
  - Localization uncertainty consideration

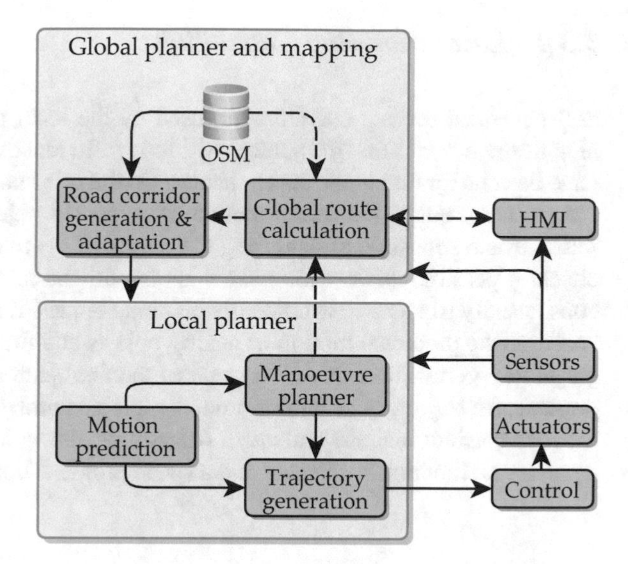

Fig. 3.4 Functional components of the architecture

- Local planning functionalities:
  - Risk estimation and motion prediction of detected dynamics obstacles.
  - Manoeuvre planner
  - Trajectory generation
- Vehicle state estimator module.
- Human-machine interface.
- Perception modules for stereo camera and LiDAR.
- Adjustments of longitudinal and lateral controllers to work with the trajectories generated by the new planning modules.

3.3.1 Global Planning Capabilities

Instead of requiring high-definition maps, the proposed approach uses low-fidelity map data to plan a global route and then automatically generate an extended data structure from OSM that enables the computation of driving corridors.

One of the main advantages of OSM is that it is an openly accessible framework to integrate mapping data. However, this may provide a possible source of data inaccuracy. These inaccuracies, together with localization measurement errors, are the main source of conflict when using OSM for automated driving. To mitigate these problems, the proposed architecture includes two complementary strategies employed in the road corridor generation and adaptation module of the architecture: a vision-based road corridor adaptation algorithm and a grid-based approach to propagate the localization uncertainty.

3.3.2 Local Planning Capabilities

Self-generated driving corridors are used by the local planner modules to finally plan the trajectories that the vehicle will follow. To that end, three different elements have been integrated in the local planner. On the one hand, (i) a **motion prediction** component predicts the future motion of perceived objects and (ii) a **manoeuvre planner** is responsible of analysing the output of the motion prediction module by checking possible spatio-temporal collisions with the current planned trajectory, and consequently trigger a new trajectory planning request if needed. On the other hand, based on the predicted motion of nearby objects and the manoeuvre request, (iii) a **trajectory generation** module computes the final path and speed profile. For that purpose, the trajectory generator produces a large number of path candidates within the road corridor that are evaluated. Afterwards, the valid candidate that minimizes a given cost function is selected and a speed profile is computed.

3.3.3 Additional Capabilities

Besides global and local planning capabilities, additional improvements on the vehicle architecture have been proposed to cope the remaining limitations identified in Sect. 3.2.3.

Firstly, a human-machine interface (HMI) has been added to the architecture for the interaction between the whole system and the vehicle occupants. This interface allows a friendly way to establish a bidirectional communication channel for both showing relevant information of the AV state to the occupants, and setting high-level commands to the vehicle.

Regarding the control module, it is worth to mention that the control approach remains the same with respect to the prior state of the architecture. However, both lateral and longitudinal controllers have been re-tuned to suit the smooth reference trajectories computed by the new local planner.

Moreover, a vehicle state estimator module has been developed to increase the reliability of vehicle pose even at low speeds. A better state estimation implies a more reliable vehicle behaviour when performing automated manoeuvres.

Since the new decision-making architecture needs to get data from the environment, new software modules to acquire data from LiDAR and camera sensors have been developed.

A full description and implementation details of each of the new developed and adapted modules is provided in section 7.2.

References

1. Artuñedo A, Godoy J, Villagra J (2019) A decision-making architecture for automated driving without detailed prior maps. In: 2019 IEEE intelligent vehicles symposium (IV) (Accepted). IEEE
2. Godoy J (2013) Arquitectura de control para la conducción autónoma de vehıculos en entornos urbanos y autovıas. PhD thesis. http://digital.csic.es/handle/10261/84925
3. Milanes V, Godoy J, Villagra J, Perez J (2011) Automated on-ramp merging system for congested traffic situations. IEEE Trans Intell Transp Syst 12(2):500–508. ISSN: 1524-9050
4. Milanes V, Onieva E, Perez J, Simo J, Gonzalez C, de Pedro T (2011) Making transport safer: A V2V-based automated emergency braking system. Transport 26(3):290–302. ISSN: 1648-4142. https://doi.org/10.3846/16484142.2011.622359. https://journals.vgtu.lt/index.php/Transport/article/view/5631
5. Milanes V, Villagra J, Perez J, Gonzalez C (20120) Low-speed longitudinal controllers for mass-produced cars: a comparative study. IEEE Trans Ind Electron 59(1):620–628. ISSN: 0278-0046. https://doi.org/10.1109/TIE.2011.2148673. http://ieeexplore.ieee.org/document/5759771/
6. Naranjo J, Bouraoui L, Garcia R, Parent M, Sotelo M (2009) Interoperable control architecture for cybercars and dual-mode cars. IEEE Trans Intell Transp Syst 10(1):146–154. ISSN: 1524-9050. https://doi.org/10.1109/TITS.2008.2011716. http://ieeexplore.ieee.org/document/4770173/
7. Naranjo J, Gonzalez C, Garcia R, DePedro T (2006) ACC+Stop&Go ManeuversWith throttle and brake fuzzy control. IEEE Trans Intelli Transp Syst 7(2):213–225. ISSN: 1524-9050. https://doi.org/10.1109/TITS.2006.874723. http://ieeexplore.ieee.org/document/1637676/
8. Perez J, Milanes V, Onieva E (2011) Cascade architecture for lateral control in autonomous vehicles. IEEE Trans Intell Transp Syst 12(1):73–82. ISSN: 1524-9050. https://doi.org/10.1109/TITS.2010.2060722. http://ieeexplore.ieee.org/document/5713840/
9. Perez J, Milanes V, Onieva E, Godoy J, Alonso J (2011) Longitudinal fuzzy control for autonomous overtaking. In: 2011 IEEE international conference on mechatronics. IEEE, pp 188–193. ISBN: 978-1-61284-982-9. https://doi.org/10.1109/ICMECH.2011.5971279. http://ieeexplore.ieee.org/document/5971279/
10. Thrun S, Montemerlo M, Dahlkamp H, Stavens D, Aron A, Diebel J, Fong P, Gale J, Halpenny M, Hoffmann G, Lau K, Oakley C, Palatucci M, Pratt V, Stang P, Strohband S, Dupont C, Jendrossek L-E, Koelen C, Markey C, Rummel C, van Niekerk J, Jensen E, Alessandrini P, Bradski G, Davies B, Ettinger S, Kaehler A, Nefian A, Mahoney P (2006) Stanley: The robot that won the DARPA grand challenge. J Field Robot 23(9):661–692. https://doi.org/10.1002/rob.20147

Chapter 4
Global Planning and Mapping

4.1 Introduction

The modules that deal with high level navigation of the vehicle given the current vehicle location and the destination are grouped in the **global planning and mapping** section of the decision-making architecture as shown in Fig. 4.4. The proposed approach relies on OSM data to firstly use a global planner to find a route that reach a destination point set through the HMI. After that, it is computed a road corridor in which the vehicle will be able to drive. As described in Sect. 3.3, the main drawback of OSM is its possible data inaccuracy. Nevertheless, a vision-based road corridor adaptation algorithm allows an on-line drift adjustment of the corridor boundaries. In addition, a probabilistic grid-based approach is also added to this architecture section in order to take into account the uncertainty of the vehicle localization.

The global route calculator is integrated with an automatic road corridor generation algorithm that receives the computed global route as input. This functional module is in turn composed of different functionalities to adapt the road corridor and to consider the uncertainty of the vehicle localization as depicted in Fig. 4.1.

The proposed vision-based algorithm for road corridor adaptation is able to provide an indicator of the success of the adaptation ("adaptation state" in Fig. 4.1). In this way, if the road corridor section that is in the field of view of the camera has been successfully adapted, the localization uncertainty is only propagated in the remaining corridor. In cases where the adaptation fails, the propagation is carried out in the whole road corridor.

Further details about the functionalities of the **global planner and mapping** section are described in the following sections: Firstly, Sect. 4.2 states some assumptions considered in the contributions of this chapter. In Sect. 4.3 the global route calculation is described. Section 4.4 addresses the road corridor generation from

A. Artuñedo, *Decision-making Strategies for Automated Driving in Urban Environments*, Springer Theses, https://doi.org/10.1007/978-3-030-45905-5_4

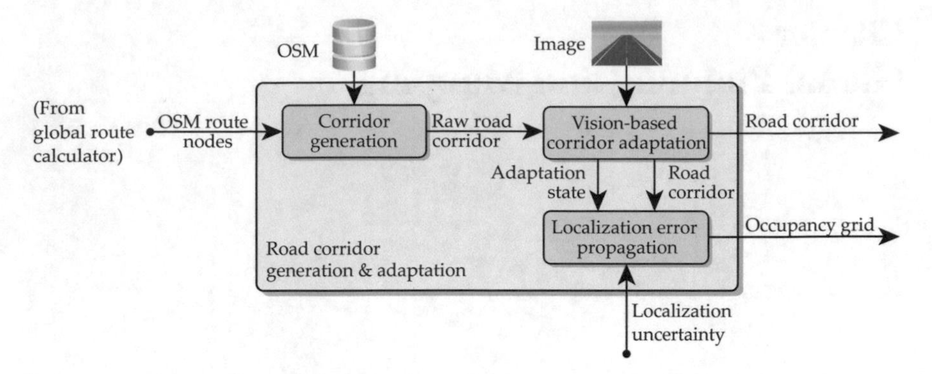

Fig. 4.1 Road corridor generation, vision-based adaptation and localization uncertainty propagation

OSM raw data. In Sect. 4.5, the vision-based road corridor adaptation algorithm is described. Finally, Sect. 4.6 introduces a probabilistic algorithm to take into consideration and propagate the localization uncertainty when using the road generated road corridors.

4.2 Assumptions

This section presents some assumptions made in the methods introduced in this chapter regarding maps, image-acquisition and localization:

Maps The road corridor generation algorithm is fed by the OSM database. The topological information is assumed to be available in all the expected application areas: urban, peri-urban, road and motorways environments. The OSM data accuracy is assumed to be high, although it is also proposed a vision-based algorithm to adapt it when the localization is good. Thus, it is considered a maximum deviation of 1 m of the map data with respect to the real ways.

Image acquisition The image positioning error of acquired images derived from camera placement and calibration is assumed to be sufficiently low to be negligible. Therefore a maximum camera positioning and angular drifts of 2 mm and 0.5° are respectively considered. Moreover, the image data needed for vision-based road corridor adaptation algorithm is assumed to be acquired at the expected rate.

4.3 Global Planner

The goal of the global planner module is to provide a high level route based on the selected destination through the HMI module and the initial vehicle location.

The global planner relies on a local instance of the Open Source Routing Machine (OSRM). This tool is based on contraction hierarchies [7], which provide on fast routing times and makes it widely accepted by the OSM community. Given the start and end points coordinates, the route is returned as a list of OSM nodes ids, which combined with the expanded structure of the nodes, is translated into a succession of segments to be joined. For simplicity of this work and attending to the European regulation, it is assumed that on the ideal scenario the vehicle travels on the right-most lane on multi-lane roads.

The map representation data used by the global planner is extracted from OSM. However, this data structure has been extended with an extra layer to automatically add information related to the state of the nodes and ways, thus allowing the modification of the travel costs of each way. This extra layer has been added keeping in mind the local dynamic map (LDM) concept that is being standardized [9] to storage environment information in different layers according their updating frequency from type 1, permanent static data, to type 4, highly dynamic data, as shown in Fig. 4.2.

Although the LDM standardization is focused on information sharing for cooperative ITS applications using V2X communications, the proposed data structure used by the architecture does not relies on external information services. Instead, it provides a way in which the different elements of the architecture, deployed in

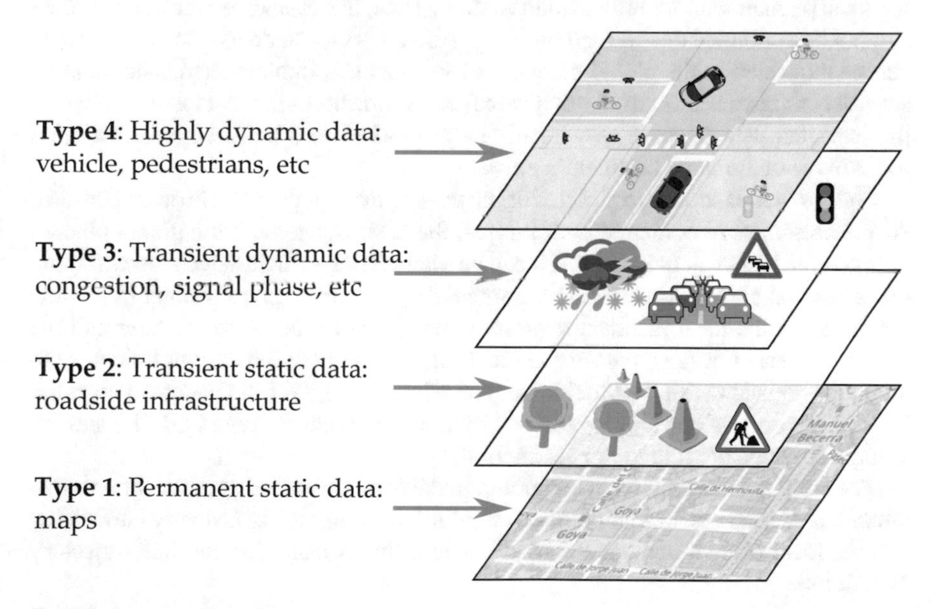

Fig. 4.2 Four layers of LDM

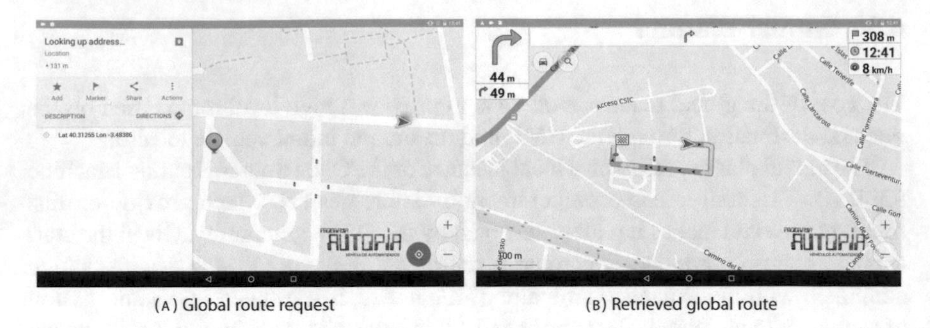

(A) Global route request (B) Retrieved global route

Fig. 4.3 HMI screenshots when requesting and retrieving a global route

the on-board vehicle computer, can provide or retrieve the required data from it. In this way, the extended data structure enables the global planner to take into account updated information about the state of the road that comes from other architecture components (such as manoeuvre planner) or even external information services using V2X communications. In this way, the routing algorithm can for instance avoid routes that pass though blocked ways.

The route calculated by the global planner is composed of a list of nodes that serve as input for the road corridor generation algorithm described in the following section. Figure 4.3A, B show two screenshots of the HMI during the request of a global route to the implemented global planner and the retrieved route, respectively.

One of the main functionalities of the global planner is its ability to interact with the local planner without human intervention. Thus, in cases where the current route can not be continued due to road blockage (road works, accidents, etc.) a new route can be requested to the global planner. To validate this implemented functionality, several tests have been carried out in a real environment. In the real testing scenario, the expected vehicle route was intentionally blocked in a point where a different route to reach the same destination exists.

Figure 4.4A, B show two pictures of the front camera of the vehicle during the test. At the bottom right corner of both figures, the HMI image with the global planner output route is shown. In Fig. 4.4A, it can be seen that the vehicle detects several static obstacles that are blocking the lane. Immediately, the manoeuvre planner (a module integrated in the local planner) requests a new route to the global planner and the trajectory generator (also part of the local planner) shorten the current trajectory to avoid the collision with the blocking object while waiting for the new road corridor. Finally, the driving corridor is received by the local planner (see Fig. 4.4B) and the vehicle continue the ride to the destination.

The global planner has been integrated in the decision-making architecture so that when a new road corridor is received, the initialization task is instantly carried out and the local planner updates the road corridor that is using for the final trajectory generation.

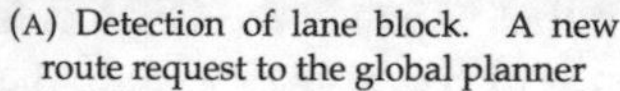

(A) Detection of lane block. A new route request to the global planner

(B) The new road corridor is received

Fig. 4.4 Frontal vehicle image and HMI screenshots when a static object that is blocking the lane is detected

4.4 Road Corridor Generation from Low Fidelity Maps

Once the high level route is defined by the global planner, the navigable space available through the route is computed. This route only consists of a list of ways and nodes through which the vehicle must drive to reach its destination. However, the OSM data structure contains useful information about each way: road width, number of lanes, driving directions, etc.

In order to obtain a driving corridor, a more refined curve representation for various lane shapes is required. As proposed in [6] or [2], the road continuity and the introduction of virtual lanes in specific situations such as road junction or exits can be taken into consideration. In this section, a solution to overcome these difficulties is proposed, providing contributions to the state of the art in the following aspects:

- A novel data structure is proposed to enhance the geometry accuracy of OSM maps, while preserving its topological structure in accordance with the real road network topology, and therefore keeping its efficiency and usability. To that end, a lane-level map will be generated guaranteeing the global continuity both in location and tangent orientation (G^1 continuity) at every node through the use of Bézier curves and a systematic procedure to softly concatenate them. This feature will be applied also to precise virtual lanes in complex roundabouts and/or intersections, defining two smooth profiles for left and right lane margins.
- A novel mechanism to automatically generate driving corridors that can be used in automated vehicles to (i) constrain the perception and localization systems to the surrounding lane(s), and (ii) to make the right decision in complex urban scenarios, and therefore to properly plan the motion to be adopted by the vehicle.

4.4.1 Road Corridor Generation Algorithm Overview

OpenStreeMap (OSM) [3] implements a topological data structure, where roads are represented as sets of nodes connected by straight segments. Hence, there is no geometric information available in the map database in order to represent curve features on roads, specially when the final goal is to define the navigable space for a vehicle.

This section proposes an automatic procedure that expands the OSM definition to generate a better approximation to the real shape of roads. To that end, the adjacency among the nodes is analysed for a given area, identifying and classifying all road junctions in order to generate an efficient and accurate polynomial-based map representation using Bézier curves [1].

In spite of being firstly used for graphic design and computer graphics, Bézier curves present some properties that have encouraged their use on other areas such as path planning or path smoothing. Some interesting properties for this work are:

- Convex hull property i.e. curve lies within the area defined by the control points.
- The derivative of a Bézier curve is, in fact, another Bézier curve. This means no derivative discontinuities.
- Tangent direction at the beginning and end of the curve is defined by $\overrightarrow{P_0P_1}$ and $\overrightarrow{P_{n-1}P_n}$, respectively, where P_0 and P_n are the start and end points of the curve, respectively.

Depending on the polynomial degree, it would be possible to define entire roads using only a Bézier curve for each one. However, the higher the degree, higher the complexity of the curve in terms of computation time and control points adjustment. Therefore, the common approach in computer graphics is to define complex paths by concatenating quadratic and cubic curves, taking into consideration the continuity constraints.

Geometrically, the continuity at the junction of two curves is expressed in terms of G^k, where k is the number of continuous derivatives at the junction. For example, G^1 refers only to the continuity of the tangent direction, while G^2 includes the continuity of the curvature.

In view of all the above, the first task of this algorithm is to modify the map description, replacing the straight segments between two consecutive nodes by cubic Bézier curves and automatically adjusting the control points for fitting roads. This is done by following a two-stages procedure: Node expansion and Bézier adjustment.

A detailed description of each stage is presented on the following subsections. Moreover, for a better comprehension of the algorithm and methods implemented, the next list summarizes the main concepts and notations used along this section. Part of them are inherited from the OSM data structure for map storage and definition and other have been created for this implementation:

Tags are key-value pairs used to describe specific features about the object they are attached.

Nodes define single points on the earth surface using WGS84 format. On this work, nodes' data structure has been modified for including further information about neighbour nodes, connecting segments and junctions/links traversing it. Along the next sections, nodes are noted as N_i, where i is the node unique id defined by OSM.

Ways are polylines representing different map features as ordered list of nodes $(N_1, N_2, ..., N_n)$. The OSM ontology defines several types of features that can be encoded as ways, but for this work only those marked with the key *highway* are considered.

Segments is one of the introduced concepts in this work. A segment refers to the road section that joins two adjacent nodes N_a and N_b. Each instance S_j includes inherited way information about the road such as number of lanes, traffic direction and road width. Along this section, segments are sometimes mentioned as *Straight* or *Bézier* segments, referring to the straight definition used by OSM or the polynomial representation defined in this section.

Links are objects describing the junction among two segments S_a and S_b sharing a common node N_l. Links metadata includes joint features such as junction angle, angle bisector, segments id, hardness and link continuity.

4.4.1.1 Node Expansion

On the first stage, the raw continuity between the segments traversing each node is determined. This concept refers to the hardness of joining two straight segments, isolated from the rest of the map. Thus, in cases such as a highway where the road is almost-straight, the segments joint smoothly with low changes on path angle and curvature. Nevertheless, in urban environments the road is curvier so the junctions are sharper.

Having the OSM map for a given area, the goal of this stage is to analyse the adjacency among the nodes in order to estimate the raw continuity of the road segments. To that end, the area is first explored way by way, creating a segment instance S_j for each pair of nodes N_a and N_b connected over a way.

Since no data about the road geometry is stored on OSM database, segments are initialized as straight Bézier curves, where the first and last control points (P_0^j and P_3^j) correspond to N_a and N_b, while the second and third points (P_1^j and P_2^j) are equally distributed between both nodes (see Fig. 4.5).

Relevant information such as traffic direction, number of lanes and road width is transferred from ways into segments, thus facilitating future adaptation of road sections individually. Moreover, the basic data structure defined by OSM for nodes is modified to store the neighbours nodes, connecting segments and raw continuity

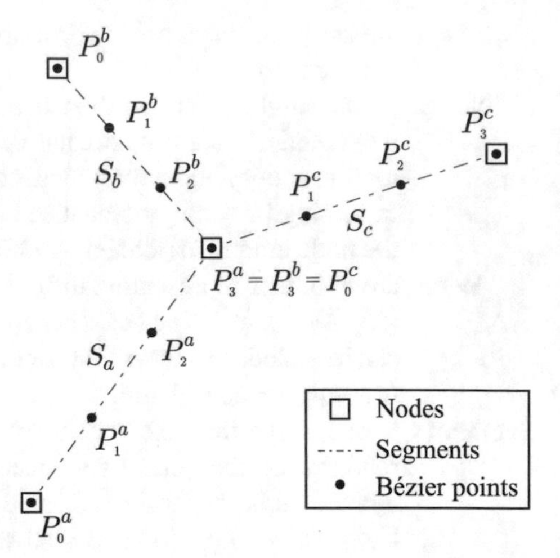

Fig. 4.5 Example of Segments with Bézier control points

among traversing segments. In this way, the algorithm is able to analyse each node taking into account all the possible connections.

Once ways have been processed and the information has been loaded into nodes, the area is re-explored node by node in order to determine the continuity among road segments. To that end, a link instance L_k is created for each pair of segments traversing a node. Thus, for a node N_i with n_v^i neighbours, the number of possible links is

$$n_l^i = \binom{n_v^i}{2} \quad (4.1)$$

A hardness value h_k describes the continuity at each junction, being $h_k = soft$ for those links possibly representing a smooth continuous road, while $h_k = hard$ for those that definitively do not. This value is set according to the link shape, taking into account the length, width and joining angle of the segments.

The selection of these features is the result of analysing the shape of several segment junctions at continuous and discontinuous scenarios such as highways, urban roads, mergings, intersections and roundabouts. From this review it was found that continuous roads are usually described by short segments and small changes on road tangents, being the opposite for discontinuous roads.

Let's consider a link L_k joining the segments $S_a = \overline{N_0 N_1}$ and $S_b = \overline{N_0 N_2}$ at the node N_0 (see Fig. 4.6). First, the angle α_k is extracted from the vector representation of the segments. If the angle is acute, L_k has a sharp shape and therefore segments are discontinuous at the junction, then $h_k = hard$. On the other hand, if the angle is obtuse the value of h_k depends also on the length and width of the segments.

As a way to simplify the analysis of all the possible scenarios, the relation among length, width and angle of the segments is expressed as two single points: triangle centroid C_k and border intersection point I_k. On the one hand, the triangle centroid

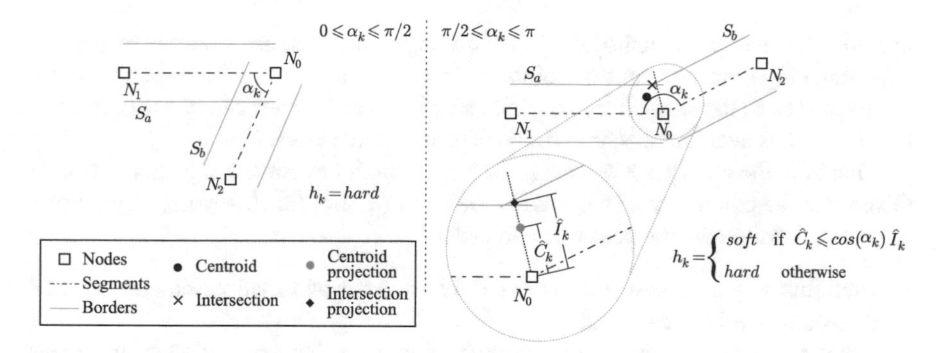

Fig. 4.6 Link hardness calculation

relies only on the position of the links nodes and therefore, the segments length and angle, while on the other hand, the intersection on the inner side of the angle relies only on the segments width and angle.

Both the centroid and the intersection point are projected over the angle bisector $\hat{m}_k$, resulting in two points whose distance to N_0 is $\hat{C}_k$ and $\hat{I}_k$, respectively. Finally, for the link to be considered $soft$ the following condition must be satisfied:

$$\hat{C}_k \leqslant \kappa \; cos(\alpha_k) \; \hat{I}_k \tag{4.2}$$

where κ is implemented as a adjustable threshold for link classification and segment modification. On this work, hardness is calculated using $\kappa = 2$, which was set experimentally after analysing the results for different maps of urban and semi-urban environments.

4.4.2 Bézier Adjustment

The second stage adjusts the control points of Bézier segments according to the fittest junction involving them. To that end, node junctions are sorted from softer to harder following several criteria, adjusting the segments in the same order. As final output, this stage produces a map where roads' centrelines have been automatically modified and better describe the real shape of the roads.

Once the nodes have been expanded and the raw continuity has been determined for all the possible links, there is enough data for adjusting the shape of the segments so they better fit the real roads. This is done by modifying the Bézier control points of the segments according to the information provided by the soft links at each node. The reason only soft links are considered on this step is because they represent the junctions were, according to our criteria, the road can be continuous.

Node by node, the proposed algorithm iterates over the soft links, selecting a target link L_t at a time for adjusting the segments associated to it. Since links only provide information about the junction of the segments, these are modified by halves, adjusting only their two control points closest to the junction. Moreover, segments are flagged once they have been modified so no further changes can be introduced at that node, thus avoiding possible overwrites among the links.

That said, the criteria for selecting the target link is key for the adjustment result. Given that the continuity of the road is the main priority for this work, target links are selected following the next rules, in order:

1. Both link segments have the same number of lanes and road width and none of them is flagged as processed.
2. Segments does not have the same number of lanes but none of them has been processed.
3. At least one of the segments remains unprocessed.

In case any of these criterion returns more than one link, those with no conflict on the associated segments are processed first. For those sharing a segment S_j, the best option is selected as the link whose angle α_t is closest to the junction angle at the opposite side of S_j. By doing so the changes in road tangent and curvature are minimized. If the segment S_j has not yet been processed at the opposite side, the adjustment at the current node is paused until angle information is available at the neighbour node.

The procedure for adjusting the control points depends of the rule triggered for the target link at the node N_i. Given a target link L_t, joining segments S_a and S_b, let's define Q_0^j and Q_1^j as the two control points of S_j closest to N_i, so either

$$Q_0^j = P_0^j \quad , \quad Q_1^j = P_1^j$$

or

$$Q_0^j = P_3^j \quad , \quad Q_1^j = P_2^j$$

and knowing $Q_0^j = N_i$ for $j = a, b$ due to segment initialization.

For the first rule to trigger, both segments must have the same road width and number of lanes, so no modification is done over the end points Q_0^A and Q_0^B since the segments' centrelines join at the same point. As for Q_1^a and Q_1^b, they are set over the perpendicular to α_t bisector $\hat{m}_t$, at the same distance d_t^1 from Q_0^a and Q_0^b, respectively. This is done so the resulting path is continuous in G^1. Parameter d_t^1 is calculated as

$$d_t^1 = min(|S_a|, |S_b|)/3 \tag{4.3}$$

In case segments do not have the same number of lanes (second criterion), it is necessary to adjust the narrower segment S_n so it joins the wider segment S_w on the correct lane. To that end, Q_0^n is moved along $\hat{m}_t$ a distance d_t^0, which value depends on the target lane. The target lane is extracted comparing the number of lanes of

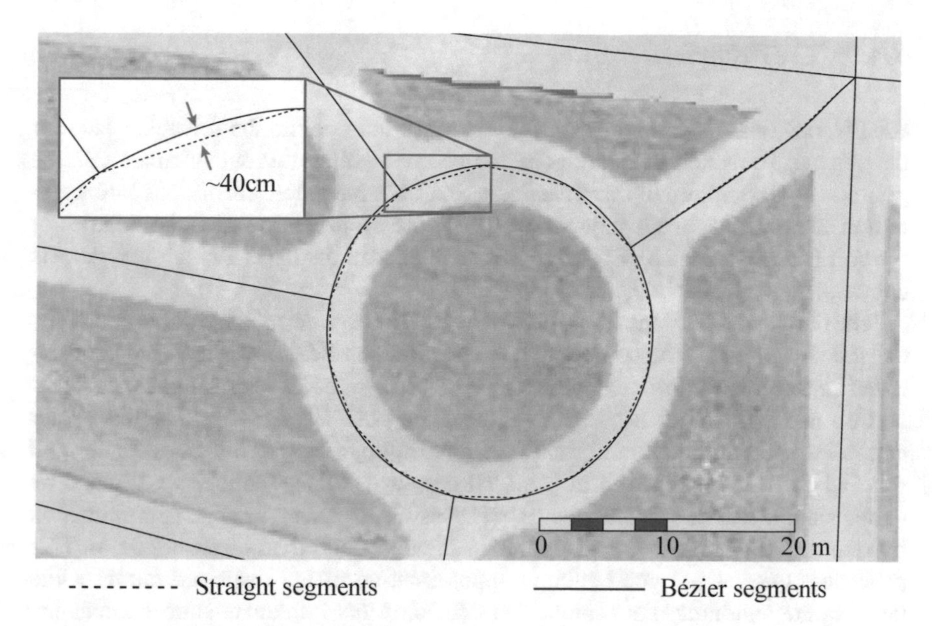

Fig. 4.7 Comparison between straight and Bézier segments on a roundabout

other nodes on the same node. Once Q_0^n is set, Q_1^a and Q_1^b are adjusted by the same procedure previously described for rule 1.

Despite the displacement applied over Q_0^n, which means Q_0^a and Q_0^b are different points, this rule also generates a junction with G^1 continuity. This is because at lane level, the joining point and road tangent are equal for both segments.

Finally, the third rule processes any remaining soft link if at least one of the segments have not been adjusted. In this case, the algorithm calculates the adjustment of Q_0^j and Q_1^j as already described for prior rules, but only modifies the segment that remains unprocessed. As only one of the segments is modified, G^1 cannot be satisfied for these kind of links. However, by adjusting the tangent at the end point, a smoother junction is achieve.

As result for this stage, the straight centrelines defining the roads have been replaced by adjusted Bézier curves that better approximate to the real paths. This improvement on the map structure is well appreciated by comparing both map definitions at curvy roads, as the roundabout shown in Fig. 4.7. In the illustrated scenario, the difference between segments at middle points is around 40 cm, which is a considerable error when using the map for path planning. For bigger roundabouts, differences up to 70 cm have been found.

4.4.3 Corridor Generation

Finally, once the map is adjusted, the final task is to define the drivable space for the vehicle. To that end, the third and final stage combines a route planner with the polynomial description of the roads, generating the navigable corridor as a concatenation of Bézier segments along a planned route. Both left and right boundaries of the corridor are defined by Bézier curves, generated as displacement of the segments centrelines.

The third and final stage of the proposed algorithm comprises the definition of the drivable space for a vehicle. To that end, the modified Bézier segments are concatenated along a planned route, thus generating the boundaries of a navigable corridor.

Left and right borders of each segment are obtained by offsetting the Bézier centreline according to its information about traffic direction, number of lanes and road width. As is mentioned in [4], a Bézier curve offset cannot be described by another single Bézier curve. However, it is possible to generate a good approximation by reducing the curve to a collection of sub-curves and then offsetting them. This procedure generates a good enough approximation of the displaced curve, while keeping the continuity constrains. The details of this implementation exceeds the scope of this work, but the reader can refer to [4] for more details about this method.

Having the borders for each segment, the navigation corridor is built taking into account the continuity of the links joining the segments. For those having already G^1 continuity at lane level, the borders are concatenated as they are defined. On the other hand, for those links classified as *hard* or without G^1 continuity, a joining section must be added in order to guarantee the G^1 continuity of all the boundaries.

Joining sections are auxiliary Bézier segments built by the algorithm from the tangent information of the segments to join. In this sense, the biggest one of the boundaries is generated and then the opposite side created as an offset of the first.

Let's consider two discontinuous consecutive segments S_a and S_b, whose left and right boundaries are $B_a^l(t)$, $B_b^l(t)$, $B_a^r(t)$ and $B_b^r(t)$, respectively. The first step is to find the boundaries intersection points at each side (I^l and I^r) in terms of the parameter t, such that

$$I^l = B_a^l(t_a^l) = B_b^l(t_b^l) \quad \text{and} \quad I^r = B_a^r(t_a^r) = D_b^r(t_b^r) \tag{4.4}$$

In case no intersection exists at the side s, then

$$t_a^s = 1 \quad \text{and} \quad t_b^s = 0 \tag{4.5}$$

considering that parameter t increases in the direction of the route. Once the intersection points are found, the algorithm selects the side w where segment a is trimmest for setting the reference t parameter for each segment. Otherwise the generated curve could not be offset for creating the opposite border. This traduces in

$$t_a = min(t_a^l,\ t_a^r) \tag{4.6}$$

and

$$t_b = \begin{cases} t_b^l, & \text{for } t_a = t_a^l \\ t_b^r, & \text{for } t_a = t_a^r \end{cases} \tag{4.7}$$

Moreover, in order to reduce the sharpness of the joining section at the smallest side and avoid it collapses into a single point, both S_a and S_b are further trimmed in terms of t_a and t_b a distance d^u from the intersection. This distance is calculated in accordance to the junction angle γ_j such that

$$d_u = max(sin(\gamma_j), 0.3) \tag{4.8}$$

This last trim is reflected on the values of t_A and t_B as $\hat{t_A}$ and $\hat{t_B}$, respectively. Finally, the control points of the auxiliary Bézier B^u are

$$\begin{aligned} P_0^w &= B_a^w(\hat{t_a}) \\ P_1^w &= p_0 + d_u\ \dot{B}_a(\hat{t_a}) \\ P_2^w &= p_3 - d_u\ \dot{B}_b(\hat{t_b}) \\ P_3^w &= B_b^w(\hat{t_b}) \end{aligned} \tag{4.9}$$

and the opposite border is generated as an offset curve from this one.

4.4.4 *Validation and Results of the Road Corridor Generation Algorithm*

The proposed algorithm for map modification and corridor generation has been implemented and applied to the surrounding area of the Centre for Automation and Robotics facilities in Arganda del Rey (Madrid, Spain). The chosen area includes both urban and interurban roads, with numerous intersections, roundabouts and merging roads. The obtained results are presented in two related subsections: (i) Comparison of map representations and (ii) Navigation corridor generation.

4.4.4.1 Map Representation

In order to demonstrate the feasibility of the proposed Bézier-based map, a comparison to the original map is presented in an scenario that included a entrance and exit of a roundabout. Figure 4.8 shows the superposition of both straight and Bézier map representations over an aerial image of the real roads.

As can be seen on this figure, the proposed road representation based on Bézier curves results in a map that is visually more accurate when compared to the aerial image. This is clearly appreciated on the joint of one-lane road segments with two-lanes ones, in contrast to the original OSM, a smooth and continuous road boundaries

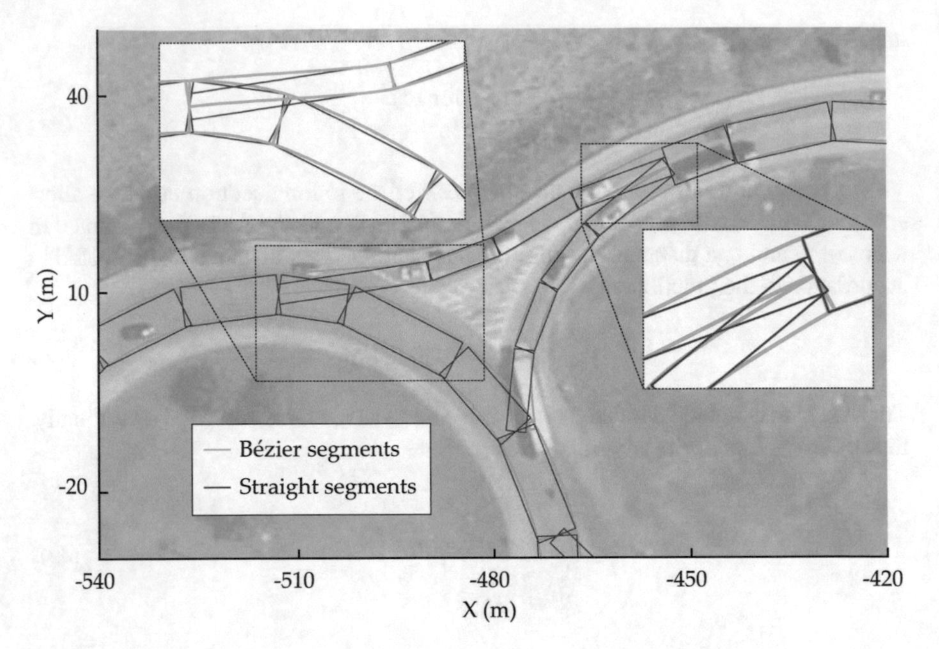

Fig. 4.8 Comparison between straight and Bézier maps in roundabout entrance & exit

shape is achieved. Moreover, as was previously commented in Sect. 4.4.2, the proposed map reduces the representation error on for curvy roads, where differences up to 70 cm have been found in the explored area.

In addition to Bézier curves, the endpoint adjustment made on narrow segments is also key for map improvement. Thanks to this modification, narrow segments join wider ones on the right point, thus reducing the lane discontinuities and overlaps found with the original map.

4.4.4.2 Navigation Corridor

Once the map of the area has been successfully adjusted and in order to validate the final stage of the algorithm, several planned routes near CAR's facilities have been introduced as input for corridor generation.

Figure 4.9 shows the details about the resulting corridor for the first route, which traverses two interurban roundabouts. As can be seen, the resulting corridor corresponds to the aerial view of the road almost perfectly, being the differences mainly caused by the image distortion and shift of the aerial pictures.

It can be appreciated that the joining sections are successfully generated in all cases, maintaining the continuity of the corridor and its tangent. Indeed, Fig. 4.10, shows the evolution of corridor tangent along all the planned route, which is continuous, so the G^1 continuity is guaranteed along the corridor.

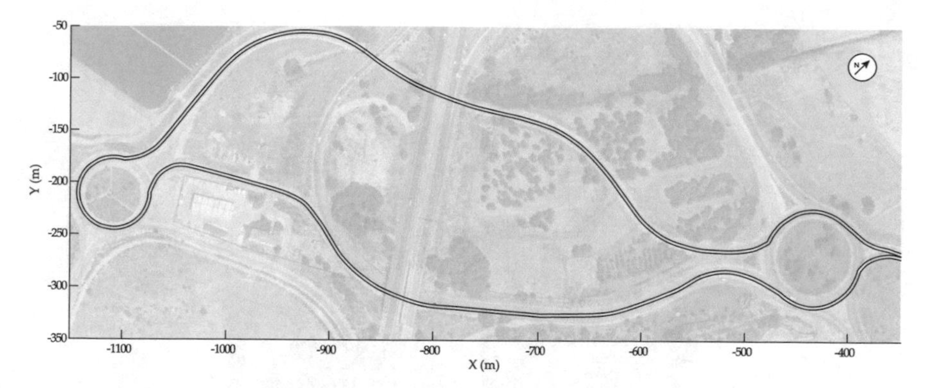

Fig. 4.9 Navigation corridor and top view image correspondence

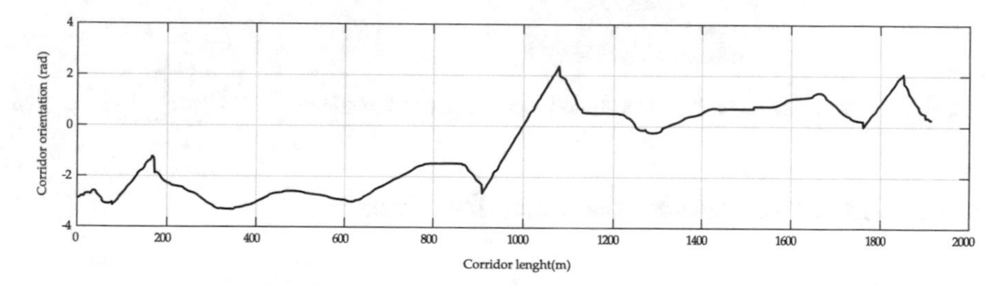

Fig. 4.10 Evolution of the corridor tangent along the planned route

4.5 Vision-Based Road Corridor Adaptation

This section introduces an algorithm for road corridor adaptation based on image processing. As mentioned in Sect. 4.4, OSM is built from information provided by the community. Therefore, although generally OSM presents accurate data, it cannot be guaranteed everywhere. Hence, the main motivation for the development of the proposed algorithm is to increase the reliability of the generated corridors in two main types of cases: (i) where the information from OSM is not accurate enough and also (ii) where the low density of OSM information does not allow the road profile to be known accurately at tight turns or intersections.

The proposed approach comprises several stages as depicted in the flow diagram of Fig. 4.11. The different stages can be grouped in three two main stages: (i) image processing and (ii) mapping & validity checking. The algorithm is designed to run every time the camera captures a new image (assuming a camera acquisition frequency of 20 Hz). Once the image is captured, the detection of road lines will be carried out in order to define the available navigable space. Then, a mapping of the environment is carried out. Finally, the results obtained are compared with the road corridor generated as described in Sect. 4.4 and the road corridor is adapted with the detected lane marks. Thus, the accuracy of the road corridor can be increased.

The following subsections provide a detailed description each algorithm stage.

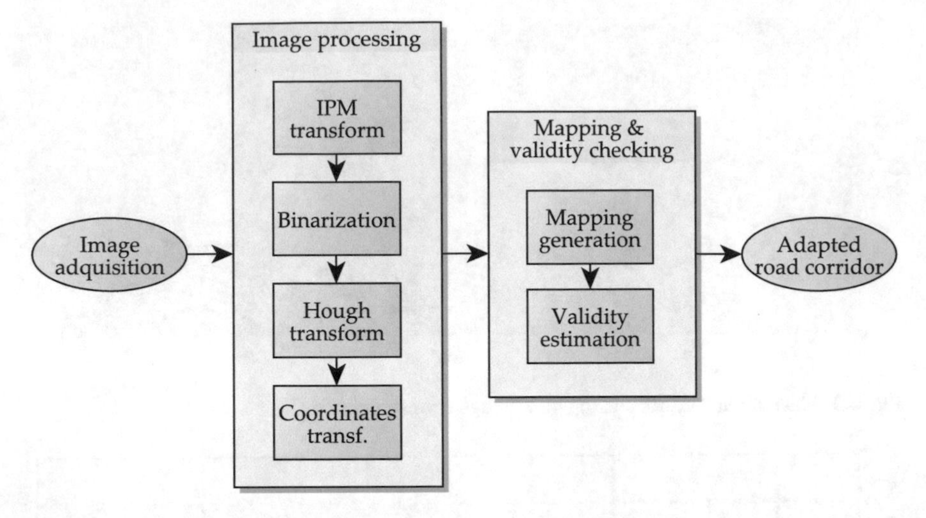

Fig. 4.11 Flow diagram of the vision-based road corridor adaptation

4.5.1 Image Processing for Lane Detection

In order to reduce the computational cost of the algorithm, the image processing is carried out on the grayscale image, thus dividing by three the processing time of a color-based approach. The following subsections describe each step of the image processing.

4.5.1.1 Inverse Perspective Mapping

After capturing an image, inverse perspective mapping (IPM) [8] is applied. Thus, the effect of perspective is removed, which is a great advantage when measuring distances in the image. In addition, the computation time is greatly reduced because the IPM focuses on the interest region of the image, excluding areas such as the hood of the car and the sky, where there is no relevant information.

An example of IPM is shown in Fig. 4.12: Fig. 4.12A shows the original image in greyscale and Fig. 4.12B shows its IPM.

In the IPM process it is assumed that the road is flat. This assumption greatly simplifies the process.

4.5.1.2 Image Binarization

Once the IPM is ready, the goal is to detect the road lines from it. To that end, a binarization process is applied in order to remove non-relevant information from the

(A) Acquired image in greyscale (B) Inverse Perspective Mapping

Fig. 4.12 Inverse Perspective Mapping application

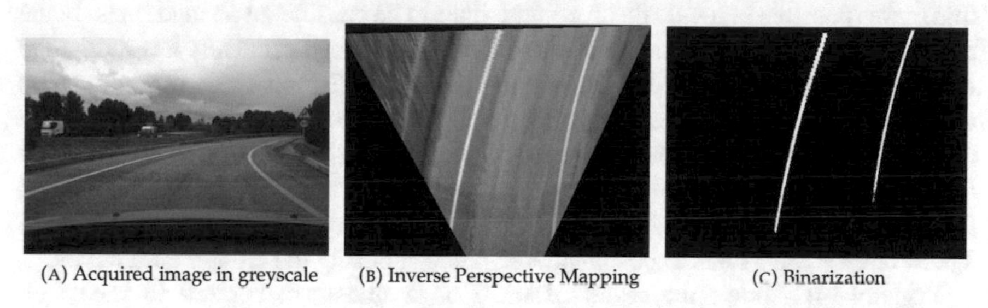

(A) Acquired image in greyscale (B) Inverse Perspective Mapping (C) Binarization

Fig. 4.13 Hough transform example

image. This process will facilitate a further detection of the lines using through the Hough transform.

In order to increase the robustness of the line detection, the binarization is based of two thresholds that are applied to different parts of the image:

Permissive threshold (th_{pr}) This threshold is applied to the surroundings of the expected positioning of the lines in the new frame.

Strict threshold (th_{st}) This stricter threshold is applied to the rest of the image.

Both thresholds are obtained from the mean of the intensity of the greyscale image (p_i) to define the final applied values as expressed in Eqs. (4.10) and (4.11).

$$th_{pr} = w_{pr} \cdot \overline{p_i} \tag{4.10}$$

$$th_{st} = w_{st} \cdot \overline{p_i} \tag{4.11}$$

In this manner it is possible to reduce the noise that may exist caused by unexpected situations such as dazzles, very clear zones in the asphalt, occlusions of the lines by other vehicles or any unforeseen object that may exist in the road.

Figure 4.13 shows an example of image binarization using the proposed method with $w_{pr} = 0.5$ and $w_{st} = 0.7$.

In some cases such as the initial frame or the first frame after finding new lines on the road, there is no previous image to estimate where the lines may be positioned to apply the permissive and strict thresholds. In these cases an overall threshold is applied for the whole image.

4.5.1.3 Road Lines Detection

The probabilistic Hough transform [5] is applied to the results obtained in the binarization. Through this procedure it is possible to detect particular features in an image, especially lines. Moreover, it is a very robust method to work in environments with possible noise or partial information occlusions.

The Hough transform returns the start and end points of each detected line. Hence, this makes possible to locate the candidate lines to be considered as road lines. In the case of roads, although the radius of curvature of the lines is relatively large, the road lines can not be always approximated with a single line. To solve that, the parameters of the Hough function are set so that the retrieved line segments are short lines. The curve approximation turns out to be better than considering long line segments, since the smaller the length of the Hough lines, the greater the road line approximation. On the other hand, the smaller the length of the lines, the greater the noise in the result. Therefore, a tradeoff has to be considered when choosing the Hough parameters.

Figure 4.14C shows the result of the Hough transform applied to image on Fig. 4.14B, which is the IPM of Fig. 4.14A.

The goal of this step is to obtain a set of points for each road corridor boundary (left and right) such that a cubic Bézier curve can be fitted. Consequently, the start and end points of the Hough lines must be grouped.

To find out if two straight lines belong to the same segment, each straight line of the Hough output has a different identifier and checked one by one if they belong to the same line of the road. To do this, a rectangle is created whose diagonal is each Hough line with a small margin. Then, it is checked if any of the ends of the rest of the lines falls within that rectangle. If so, it is assumed that both lines belong to the same segment, so they are assigned with the same identifier.

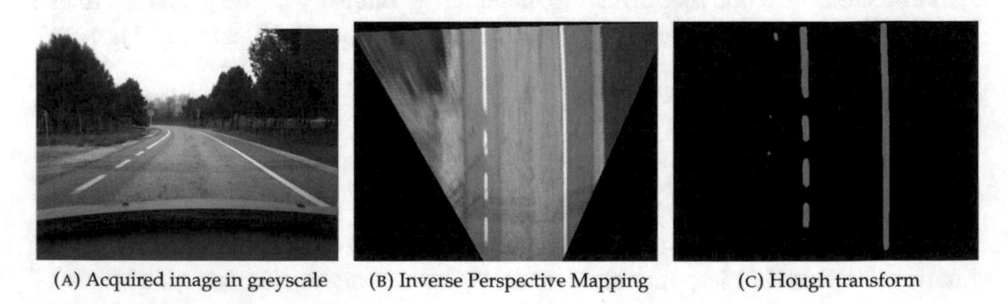

(A) Acquired image in greyscale (B) Inverse Perspective Mapping (C) Hough transform

Fig. 4.14 IPM binarization example

Once the points are assigned to the left or right boundary group, a cubic Bézier curve is approximated to the data by using the least-squares fitting method.

4.5.1.4 Coordinates Transformation

So far, the different processing stages have been using the coordinate system of the image. In order to compare the results with the previous road corridor generated from OSM, the resulting road line detection must be transformed to the same reference system. Since the road corridor uses the global UTM coordinate system, the generated road lines from the image are transformed to this reference system.

The first coordinate change is performed from the reference system of the IPM image O_i to the system O_u whose coordinate centre is placed a given distance ahead the vehicle ($d_u = 5.4$ m) with a calibration angle ($\theta_c = 0.0908$ rad) as can be seen in Fig. 4.15. The coordinates in O_i are expressed in pixels while in O_u, O'_u and O_G (the global UTM coordinate system) are expressed in meters. Therefore, it is needed to know the amount of pixels per meter. The value determined from the camera calibration is: $r_{pm} = 32.67$ pixels/m.

Since the IPM image has a resolution of 480 × 360 pixels, the conversion from O_i to O_u is carried out as follows:

$$\begin{aligned} X_u &= \frac{Y_i - 480/2}{r_{pm}} \\ Y_u &= \frac{360 - X_i}{r_{pm}} \end{aligned} \tag{4.12}$$

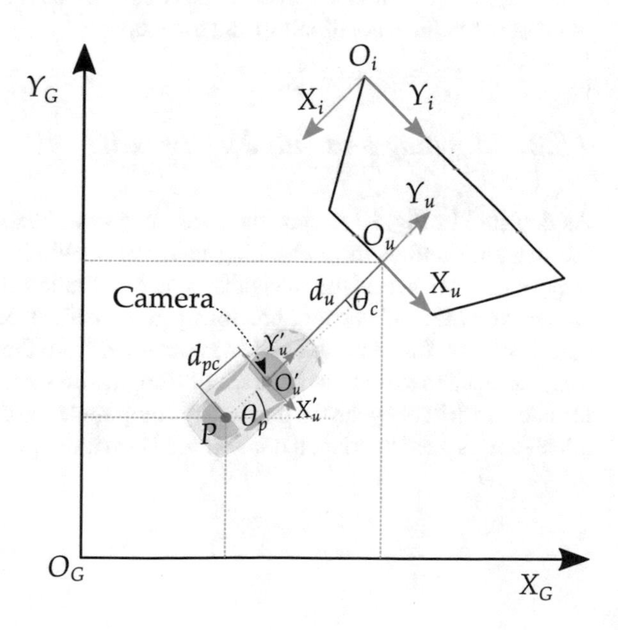

Fig. 4.15 Vehicle and camera coordinate systems

In the transformation from O_u to O'_u, only Y'_u changes:

$$X'_u = X_u \tag{4.13}$$

$$Y'_u = Y_u + d_u \tag{4.14}$$

Then, a transformation from the coordinate system O'_u to O_G is needed. A rotation to consider the camera calibration angle is performed using the following rotation matrix:

$$R(\theta_c) = \begin{pmatrix} \cos(\theta_c) & -\sin(\theta_c) \\ \sin(\theta_c) & \cos(\theta_c) \end{pmatrix} \tag{4.15}$$

After that, the position and orientation of the vehicle (X_p, Y_p, θ_p) are used. The rotation angle is given by

$$\alpha = \pi/2 - \theta_p \tag{4.16}$$

Thus, the final conversion can be carried out by using the rotation matrix defined as follows:

$$R(\alpha) = \begin{pmatrix} \cos(\alpha) & -\sin(\alpha) \\ \sin(\alpha) & \cos(\alpha) \end{pmatrix} \tag{4.17}$$

$$\begin{pmatrix} X_G \\ Y_G \end{pmatrix} = \begin{pmatrix} X_p + d_{pc}\cos(\theta_p) \\ Y_p + d_{pc}\sin(\theta_p) \end{pmatrix} + R(\alpha)\begin{pmatrix} X_u \\ Y_u \end{pmatrix} \tag{4.18}$$

where $d_{pc} = 2.25$ m is the distance between the centre of the rear axle (reference point of the vehicle positioning) and the camera.

4.5.2 Mapping and Validity Checking

As depicted in Fig. 4.11, once the road lines have been computed from the image, the mapping and validity checking task starts. Within this process, a comparison is performed between the navigable space delimited by the lines detected by the camera and the road corridor previously generated, as described in Sect. 4.4. To that end, firstly the final left an right lines obtained from the image are computed. After that, it is needed to find out if the result from the image processing is valid. Thus, if it is valid, the drift between the previously computed road corridor and the vision-based adaptation is used to adapt the initial road corridor.

4.5.2.1 Vision-Based Map Generation

The developed algorithm only runs the mapping of a new section of the road each time a new control point enters the field of view of the camera. This is to ensure that the new computed Bézier is of the right dimensions to compare and replace it with the Bézier that belongs to the original road corridor.

Once the algorithm detects that one of the new Bézier sections of the original road corridor is located in the area covered by the IPM, the line detection is triggered. After that, the new lines detection is combined with the next section of the original road corridor to obtain a navigable space from areas that escape the effective field of view of the camera.

Finally, as both sections do not usually match, a new curve is generated in order to join the vision-based map and the original road corridor. This curve section is added in order to smooth the transition between the image field of view and the next part of the road corridor that still has not been reached by the camera.

Figure 4.16 shows a schematic example of the method used to determine the new joint section. In this figure, the road section that has already been adapted through the image is shown in red. The green zone is the map obtained from the original road corridor.

Note that the normalized tangent vectors at the last control point of the adapted curve (P_3^a) and at the first control point of the next original section (P_0^o) and the distance between them are used to place the intermediate points of the joint section (P_1^j and P_2^j) as described by Eqs. (4.19–4.22). This method allows to guarantee the G^1 continuity at the extremes of the joint section since P_2^a, $P_3^a \equiv P_0^j$ and P_1^j are aligned, as are P_2^j, $P_3^j \equiv P_0^o$ and P_1^o.

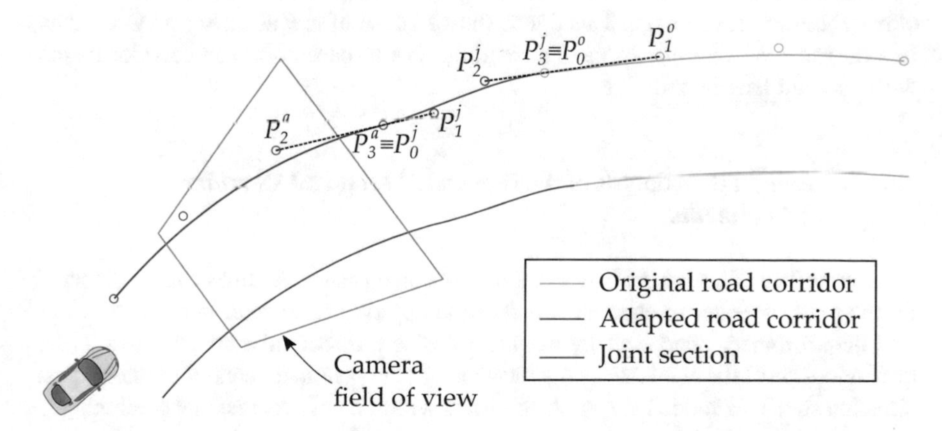

Fig. 4.16 Joint section generation between the adapted and the original road corridors

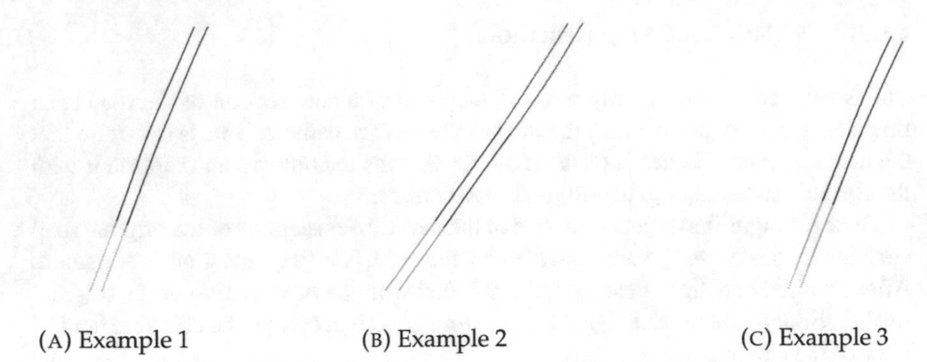

(A) Example 1 (B) Example 2 (C) Example 3

Fig. 4.17 Three examples of the vision-based road corridor adaptation

$$P_0^j = P_3^a \quad (4.19)$$

$$P_1^j = P_3^a + \frac{1}{3} L_j \mathbf{t}_3^a \quad (4.20)$$

$$P_2^j = P_0^o - \frac{1}{3} L_j \mathbf{t}_0^o \quad (4.21)$$

$$P_3^j = P_0^o \quad (4.22)$$

where $\mathbf{t}_3^a$ and $\mathbf{t}_0^o$ are the normalized tangent vectors at P_3^a and P_0^o, respectively, and L_j is the euclidean distance between P_3^a and P_0^o.

In Fig. 4.17, the joint sections computed in three different real situations are shown. Note that the transitions generated by the joint section between the adapted and the original road corridors are smooth in all cases.

A result of the raw and adapted corridors is shown in Fig. 4.18 together with the corresponding aerial image. This figure shows a case of low accuracy on OSM data in a curved road section. As can be seen, the vision-based adapted corridor match with the road lane boundaries.

4.5.2.2 Reliability Analysis of the Generated Map and Corridor Adjustments

The variability of scenarios is so great that in some cases fake lines can be detected on the road. In order to estimate whether the approximated road lines are valid or not, the aforementioned strategy has been devised to recognise major flaws in the determination of the available navigable space. If the system detects some anomalous situation such as a sudden change in the lanes width, it will assume that something is not working well and the system will report an alert. Consequently the road corridor shall not be rectified in these cases.

In order to quantify the variation of the lane width, it is assumed that the lines to be detected over the road are parallel i.e. the width of the lane is constant. Therefore, the

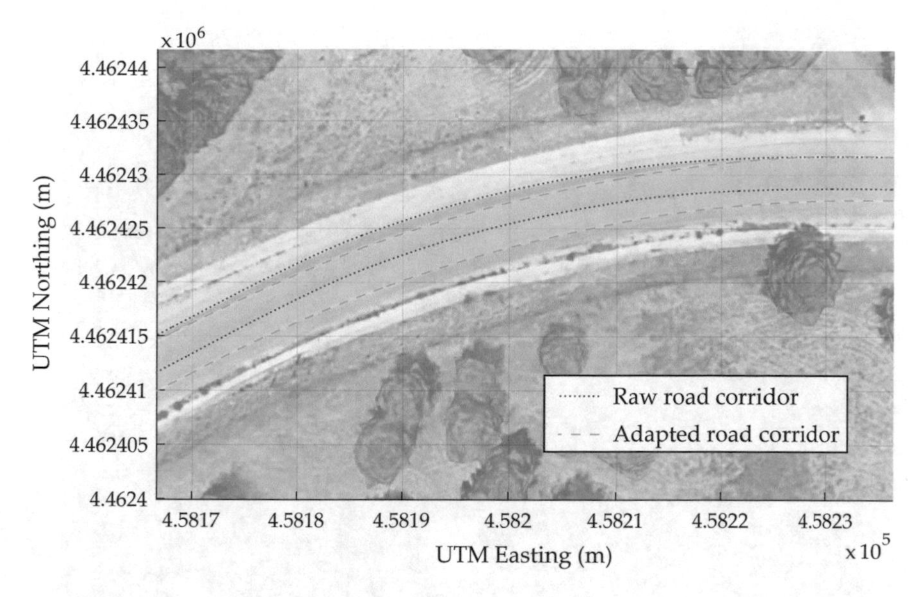

Fig. 4.18 Results of the vision-based adaptation

width of a set of points of the current detected lane is evaluated using the perpendicular distance from one Bézier curve to the other. After that, the mean of the widths is used to calculate the standard deviation of the last $n_{\sigma w}$ width mean values in order to provide temporal consistency of the indicator (σw). Finally, a threshold ($th_{\sigma w}$) is used to determine whether the result of the vision-based lines detection is valid or not.

So as to choose the values of both $n_{\sigma w}$ and $th_{\sigma w}$ a number of tests were carried out concluding that $n_{\sigma w} = 20$ and $th_{\sigma w} = 0.3$ were reasonable values to detect the algorithm faults.

Figure 4.19A, B shows the projection of the vision-based road corridor detection on the original acquired image. On the one hand, Fig. 4.19A shows a typical road where the detection algorithm seamlessly finds the road corridor in the image. Hence, the detected road corridor is shaded green as $\sigma w < th_{\sigma w}$. On the other hand, Fig. 4.19B shows an algorithm fault where the left line of the lane is not correctly detected. Nevertheless, this fault makes σw to be increased over the threshold value ($\sigma w > th_{\sigma w}$). Consequently, the fault is detected and the road corridor is shaded red.

4.6 Considering Localization Uncertainty When Using Road Corridors

Localization plays an important role in autonomous driving. In particular, when maps are used as a part of the environment understanding, a good localization with respect to the map becomes more critical. In some cases, the accuracy of localization systems

(A) The road corridor is shaded green as the detection is considered valid ($\sigma w < th_{\sigma w}$)

(B) The road corridor is shaded red as the detection is considered invalid ($\sigma w > th_{\sigma w}$)

Fig. 4.19 Result of the projection of the vision-based road corridors over the original images

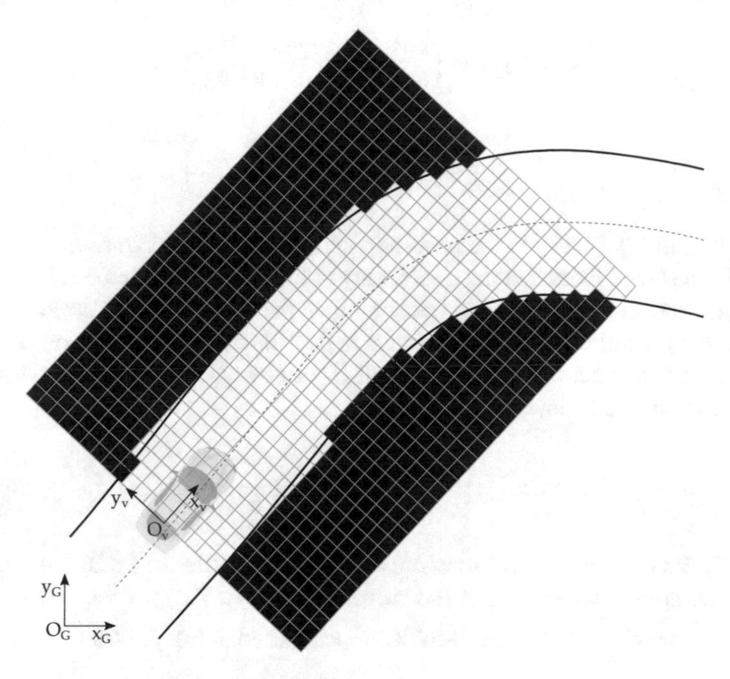

Fig. 4.20 Road corridor rasterization over the grid

can be low by design or even can drop depending on the environment (e.g. GPS-based localization systems when there are reflections in urban canyons or satellites occlusions, or in cloudy scenarios). In these situations, the localization uncertainty can be taken into consideration to increase the system reliability.

To consider the localization uncertainty, a similar approach to the one defined in [10] is used. This method requires the environment data to be represented in an occupancy grid, which provides a way to represent probabilistic information generated from different sensors measurements taking into account their noise and uncertainty. Then vehicle pose uncertainty is propagated along the occupancy grid, obtaining the occupancy probability of each cell of the grid.

The first step is to represent the map over the grid. In our case, the information of the map is composed of a set of Bézier curves that define the left and right boundaries of the navigable space. In order to set the occupancy of each Bézier segment over the grid, an extension of the Bresenham algorithm for cubic Béziers [11] is applied. After that, the free space existing inside the road corridor is filled with null occupancy probability while the rest of the grid if set as occupied (see Fig. 4.20).

Finally, the uncertainty of the vehicle pose is computed for all the initially free cells using the general approach defined in [10]. Let Cx_i and Cy_i denote the coordinates of the cell i of the grid in the frame O_v, and Vx_{O_G}, Vy_{O_G} and $V\theta_{O_G}$ denote the vehicle position and heading in the frame O_G. Then, it is transformed into the frame O_G as follows:

$$R(V\theta_{O_G}) = \begin{pmatrix} \cos(V\theta_{O_G}) & -\sin(V\theta_{O_G}) \\ \sin(V\theta_{O_G}) & \cos(V\theta_{O_G}) \end{pmatrix} \tag{4.23}$$

$$\begin{pmatrix} x_{O_G} \\ y_{O_G} \end{pmatrix} = R_{vG} \begin{pmatrix} Cx_i \\ Cy_i \end{pmatrix} + \begin{pmatrix} Vx_{O_G} \\ Vy_{O_G} \end{pmatrix} \tag{4.24}$$

where R_{vG}, in Eq. (4.23), is the rotation matrix from the vehicle frame (O_v) to the global frame (O_G). Note that the uncertainty of the position of each cell in the global frame (x_{O_G}, y_{O_G}) comes from the uncertainty in the vehicle pose in the global frame (Vx_{O_G}, Vy_{O_G}) as the position of the cells in the vehicle frame (Cx_i, Cy_i) is known.

The covariance matrix (g_i) in each cell (i) can be calculated from the estimated pose uncertainty as follows:

$$g_i(x_i, y_i) = \left(\frac{\delta T_f}{\delta V_{O_G}}\right) Q_V \left(\frac{\delta T_f}{\delta V_{O_G}}\right)^T \tag{4.25}$$

where $T_f(Vx_{O_G}, Vy_{O_G}, V\theta_{O_G})$ denotes the transformation in Eq. (4.24), Q_V represents the covariance matrix of the current pose $V_{O_G}(Vx_{O_G}, Vy_{O_G}, V\theta_{O_G})$ in the global frame and $\left(\frac{\delta T_f}{\delta V_{O_G}}\right)$ is the Jacobian as expressed in Eq. (4.26).

$$\left(\frac{\delta T_f}{\delta V_{O_G}}\right) = \begin{pmatrix} \frac{\delta T_f}{\delta Vx_{O_G}} \\ \frac{\delta T_f}{\delta Vy_{O_G}} \\ \frac{\delta T_f}{\delta V\theta_{O_G}} \end{pmatrix}^T \tag{4.26}$$

Thus, the resulting covariance matrix is shown in Eq. (4.27).

$$g_i(x_i, y_i) = \begin{pmatrix} \sigma_{xi}^2 & \rho_i \sigma_{xi} \sigma_{yi} \\ \rho_i \sigma_{xi} \sigma_{yi} & \sigma_{yi}^2 \end{pmatrix} = \begin{pmatrix} \sigma_x^2 + \sigma_\theta^2\, u(V\theta_{O_G}) & \sigma_\theta^2\, t(V\theta_{O_G}) \\ \sigma_\theta^2\, t(V\theta_{O_G}) & \sigma_y^2 + \sigma_\theta^2\, v(V\theta_{O_G}) \end{pmatrix} \tag{4.27}$$

where:

$$u(V\theta_{O_G}) = (-\sin(V\theta_{O_G})\, Cx_i - \cos(V\theta_{O_G})\, Cy_i)^2$$
$$v(V\theta_{O_G}) = (\cos(V\theta_{O_G})\, Cx_i - \sin(V\theta_{O_G})\, Cy_i)^2$$
$$t(V\theta_{O_G}) = \sin(V\theta_{O_G}) \cos(V\theta_{O_G})(Cx_i^2 - Cy_i^2) + Cx_i Cy_i (\sin(V\theta_{O_G})^2 - \cos(V\theta_{O_G})^2)$$

σ_x, σ_y and σ_θ are the given pose uncertainties of the vehicle, and $\rho_i = \frac{\sigma_\theta^2\, t(V\theta_{O_G})}{\sigma_{xi}\, \sigma_{yi}}$.

Once the covariance matrix g_i is known, a bivariate Gaussian distribution is applied to compute the probability distribution for each cell, as expressed in Eq. (4.28).

$$f_i(x_j, y_j) = \frac{\exp\left\{-\frac{1}{2(1-\rho_i^2)}\left[\left(\frac{Cx_j - Cx_i}{\sigma_{xi}}\right)^2 - 2\rho_i\left(\frac{Cx_j - Cx_i}{\sigma_{xi}}\right)\left(\frac{Cy_j - Cy_i}{\sigma_{yi}}\right) + \left(\frac{Cy_j - Cy_i}{\sigma_{yi}}\right)^2\right]\right\}}{2\pi\sigma_{xi}\sigma_{yi}\sqrt{1-\rho_i^2}} \tag{4.28}$$

To compute final occupancy probability of each cell, a 95% confidence ellipse ($\chi^2 = 5.991$) is defined from the computed covariances. Then, the final occupancy probability of each cell is calculated from the expected occupancy values and probability of the cells that fall within the ellipse, as expressed in Eq. (4.29).

$$P(x_i, y_i) = \frac{\sum_{j \in I_i} f_i(x_j, y_j) \cdot F_j}{\sum_{j \in I_i} f_i(x_j, y_j)} \tag{4.29}$$

where I_i is the set of cells that falls within the ellipse generated for cell i, j is the index of the cells inside I_i, S is the size of set I_i, $f_i(x_j, y_j)$ is the probability in cell j obtained by the Gaussian distribution in Eq. (4.28), generated for the cell i, and $F_j \in \{0, 1\}$ is the initial occupancy value of the cell j.

An schematic example of the result is shown in Fig. 4.21, where cell colours vary from white (free cells) to black (occupied cells) passing through grey tones representing different probability values. Besides, a summary of the localization propagation algorithm over the road corridor is shown in algorithm 1.

Input: Pose uncertainty ($\sigma_x, \sigma_y, \sigma_\theta$), Road corridor
Output: Occupancy probability in each cell
grid ← Left and right boundaries of the road corridor ;
grid(cells inside road corridor) ← 0 (empty) ;
grid(cells outside road corridor) ← 1 (occupied) ;
foreach $i \leftarrow 1$ ***to*** n **do**
    compute covariance matrix [Eq. (4.25)]
    compute bivariate Gaussian distribution [Eq. (4.28)]
    $grid(i) \leftarrow \frac{\sum Expected\ values}{\sum Probabilities}$ [Eq. (4.29)];
end
return *grid*;

Algorithm 1: Localization uncertainty propagation over an occupancy grid

The uncertainty propagation over the grid results in the narrowing of the road corridor, being this effect particularly pronounced when the heading uncertainty is high. The occupancy probability of a priori free cells becomes higher when the x_v coordinate is increased as can be seen in Fig. 4.21.

Figure 4.22 shows the results of uncertainty propagation using different values of σ_x, σ_y and σ_θ. In all cases the grid size is 20 × 30 meters with a grid resolution of 20 cm.

Figure 4.22A shows the road corridor rasterization over the grid, assuming negligible localization uncertainties. Comparing Fig. 4.22B, C it can be observed the effect of localization uncertainty propagation when the different between σ_x and σ_y

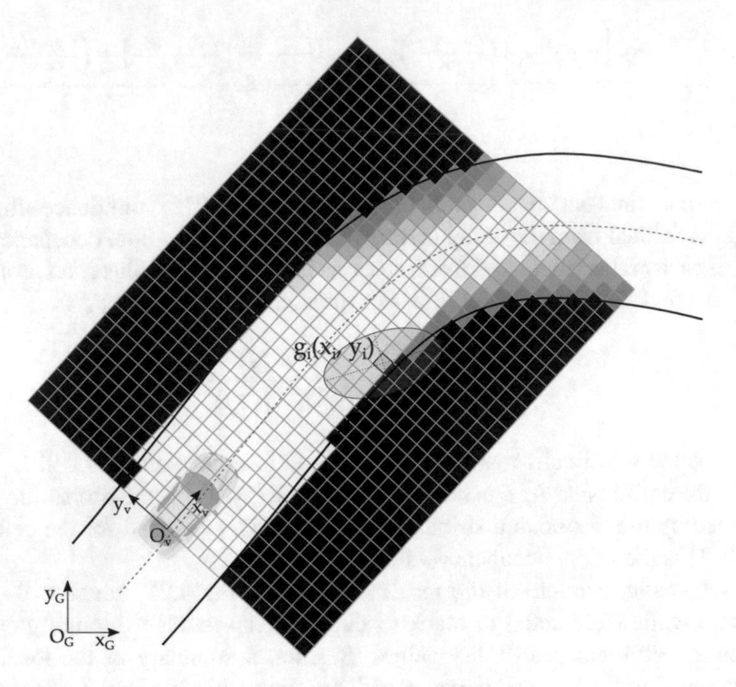

Fig. 4.21 Localization uncertainty propagation over the grid

is high. As can be seen, if the longitudinal uncertainty is high and it is low in the lateral axis, the known free space inside the road corridor is narrowed. In the opposite case (Fig. 4.22C), most of the cells at both sides of the vehicle are ensured to be unoccupied.

As can be observed in Fig. 4.22D, where $\sigma_x = \sigma_y = 0.5\ m$ and $\sigma_\theta = 0\ rad$, the most of the cells occupied by road corridor are completely free even with high uncertainties in X and Y axis. Furthermore, the case represented in Fig. 4.22E adds a small orientation uncertainty with respect to the previous one. This leads to the narrowing of the road corridor in the farthest cells from the vehicle. Thus, these examples depict the high influence of the orientation uncertainty in comparison with X and Y ones.

Finally, a case with high longitudinal, lateral and angular uncertainties is shown in Fig. 4.22F, where only the cells closest to the vehicle are guaranteed to be unoccupied.

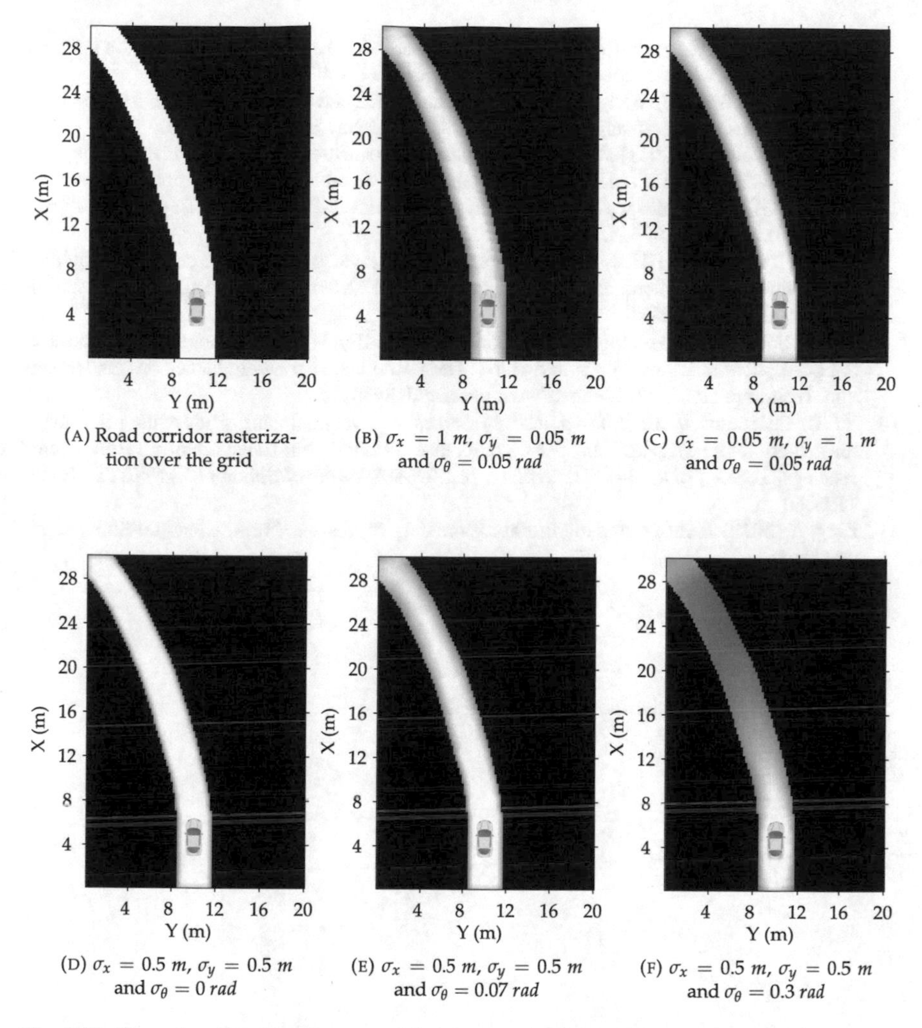

(A) Road corridor rasterization over the grid

(B) $\sigma_x = 1\ m$, $\sigma_y = 0.05\ m$ and $\sigma_\theta = 0.05\ rad$

(C) $\sigma_x = 0.05\ m$, $\sigma_y = 1\ m$ and $\sigma_\theta = 0.05\ rad$

(D) $\sigma_x = 0.5\ m$, $\sigma_y = 0.5\ m$ and $\sigma_\theta = 0\ rad$

(E) $\sigma_x = 0.5\ m$, $\sigma_y = 0.5\ m$ and $\sigma_\theta = 0.07\ rad$

(F) $\sigma_x = 0.5\ m$, $\sigma_y = 0.5\ m$ and $\sigma_\theta = 0.3\ rad$

Fig. 4.22 Examples of localization uncertainty propagation over a 20 m × 30 m occupancy grid

References

1. Godoy J, Artunedo A, Villagra J (2019) Self-generated OSM-based driving corridors IEEE Access 7:20113–20125. ISSN: 2169-3536. https://doi.org/10.1109/ACCESS.2019.2897348
2. Guo C, Meguro J-I, Kojima Y, Naito T (2014) Automatic lane-level map generation for advanced driver assistance systems using low-cost sensors. In: IEEE international conference on robotics and automation (ICRA). IEEE, pp 3975–3982
3. Haklay M, Weber P (2008) OpenStreetMap: user-generated street maps IEEE Pervas Comput 7.4:12–18. ISSN: 1536-1268. https://doi.org/10.1109/MPRV.2008.80. http://ieeexplore.ieee.org/document/4653466/
4. Kamermans M (2016) A Primer on Bézier Curves. https://pomax.github.io/bezierinfo/

5. Kiryati N, Eldar Y, Bruckstein AM (1991) A probabilistic Hough transform. Pattern Recogn. ISSN: 00313203. https://doi.org/10.1016/0031-3203(91)90073-E
6. Liu J, Cai B, Wang Y, Wang J (2013) Generating enhanced intersection maps for lane level vehicle positioning based applications. Procedia-Soc Behav Sci 96:2395–2403
7. Luxen D, Vetter C (2011) Real-time routing with OpenStreetMap data. In: Proceedings of the 19th ACM SIGSPATIAL international conference on advances in geographic information systems. GIS '11. Chicago, Illinois: ACM, pp 513–516. ISBN: 978-1-4503-1031-4. https://doi.org/10.1145/2093973.2094062
8. Mallot HA, Bülthoff HH, Little JJ, Bohrer S (1991) Inverse perspective mapping simplifies optical flow computation and obstacle detection. Biol Cybern. ISSN: 03401200. https://doi.org/10.1007/BF00201978
9. TR 102 863 - V1.1.1 - Intelligent Transport Systems (ITS); Vehicular Communications; Basic Set of Applications; Local Dynamic Map (LDM); Rationale for and guidance on standardization. Tech. rep. ETSI, 2011. https://www.etsi.org/deliver/etsi
10. Yu C, Cherfaoui V, Bonnifait P (2016) Semantic evidential lane grids with prior maps for autonomous navigation. In: IEEE conference on intelligent transportation systems, proceedings, ITSC, pp 1875–1881. ISBN: 9781509018895.https://doi.org/10.1109/ITSC.2016.7795860
11. Zingl A (2012) A rasterizing algorithm for drawing curves. In: Multimedia und Softwareentwicklung

Chapter 5
Motion Prediction and Manoeuvre Planning

5.1 Introduction

Among all traffic accidents with personal injuries in urban environments, more than the 50% occur at intersections [5]. These accidents are typically caused by the impossibility to understand the nearby environment, predict future states, and act consequently by human drivers in critical situations where the behaviour of traffic scene agents is highly unpredictable and the time to act is very low. In this sense, the anticipation to future driving scenes plays a key role in automated driving.

Because of the high variety of possible manoeuvres that a driver can perform in common urban situations, a comprehensive understanding of the current state of the nearby environment is not enough to take an optimal decision. Information about the future state of the environment is also needed to be provided to the decision system in order to quantify the risk of possible ego-vehicle manoeuvres and then act consequently. Due to the complexity of the motion prediction problem, currently there are no ADAS in production that can issue early warnings of collisions at urban intersections.

Bearing this in mind, the estimation of future movements of relevant objects in the nearby environment of the ego-vehicle is addressed by the proposed architecture using the motion prediction module. To that end, this module receives as input the perceived objects, the road corridor and additional information from maps about traffic rules; and outputs the predicted trajectory of the most hazardous dynamic objects. This module includes two different functionalities as depicted in Fig. 5.1: firstly, (i) a risk estimation algorithm [7] uses information of traffic rules from maps and the state of the perceived objects (position, orientation, speed, etc.) to select the most hazardous objects perceived by the vehicle in highly complex situations i.e. when there are many agents involved in the driving scene. Finally, (ii) the future trajectory of the selected moving objects is predicted and passed to the manoeuvre planner module to be analysed and generate an output action if needed.

A. Artuñedo, *Decision-making Strategies for Automated Driving in Urban Environments*, Springer Theses, https://doi.org/10.1007/978-3-030-45905-5_5

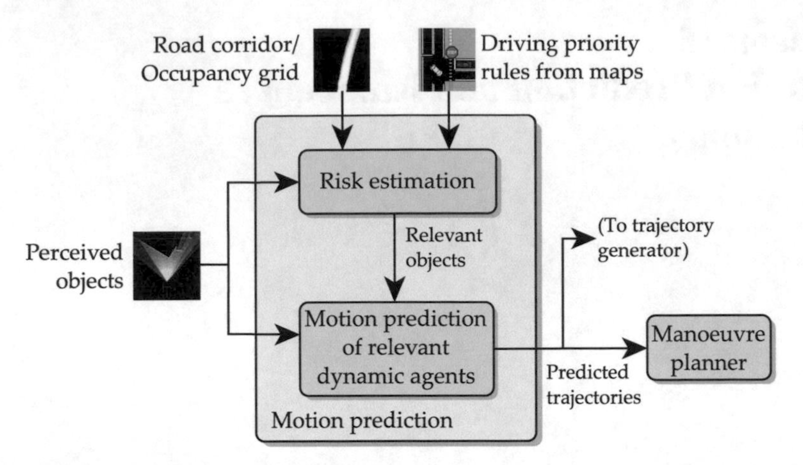

Fig. 5.1 Functional diagram of the motion prediction module

The chapter is structured as follows. Firstly, Sect. 5.2 states some assumptions considered in the contributions of this chapter. The subsequent section focuses on three related aspects of the decision-making architecture. On the one hand, the work on risk estimation is presented in Sect. 5.3. This section introduces an interaction-aware approach for risk estimation considering both longitudinal and lateral manoeuvres. Preliminary results of the risk estimation algorithm are presented for different scenarios. On the other hand, Sect. 5.4 focuses on the simplified motion prediction included in the implemented architecture. Finally, Sect. 5.5 introduces the manoeuvre planner component of the architecture, which is in charge of deciding future manoeuvres of the ego-vehicle.

5.2 Assumptions

Some assumptions are made in both methods introduced in this chapter: risk estimation algorithm and manoeuvre planner:

Maps and road corridors As in Chap. 4, the accuracy of the maps is assumed to be good enough to be used for the road corridor, that is used by the manoeuvre planner to detect static objects that could interfere with the ego-vehicle trajectory. Thus, it is considered a maximum deviation of 1 m of the map data with respect to the real ways.

Perception The manoeuvre planning algorithm proposed in this chapter uses the information provided by the perception systems to estimate the state of nearby obstacles. The inaccuracy of this information is assumed to be low enough.

Localization The uncertainty in global localization is assumed to be negligible so that it is not considered, assuming a maximum localization error of 5 cm and 1?

of heading error. Thus, the ego-localization with respect to the maps which are placed in the global coordinates frame is assumed to be good enough.

Computational resources The risk estimation algorithms are intended to provide a risk indicator in real-time decision systems. The algorithm described in this chapter needs high computation resources to provide a result in a reasonable time. For this first implementation it is assumed the availability of resources to provide the risk estimation in the expected time limit, so that the run-time of the algorithms is not considered.

Motion prediction simplification The manoeuvre planner implemented in the architecture tested in the real vehicle needs to predict the movements of the detected dynamic objects. In this first implementation it is included a simplified motion prediction algorithm that provides a good enough trajectory prediction for the motion planner tp act consequently in case of possible collisions, without needing large computational resources.

5.3 Risk Estimation in Urban Environments

Motion prediction plays a key role in autonomous vehicles. In fact, in typical urban driving scenes such as intersections, a large number of traffic agents can present a collision risk for the ego-vehicle. Indeed, it is extremely difficult to predict individually and accurately the trajectories of each agent in scenes with a large number of them, in which their behaviours have clear interactions. Besides, the limited computational resources in the vehicle could not be enough to compute complex prediction algorithms in a short time frame, the use of an interaction-aware risk estimation algorithm becomes of great interest.

As illustrated in Sect. 2.3, different approaches coexist in the literature to estimate the collision risk with other traffic agents. In this sense, this section presents a study conducted to explore the potential of intention estimation techniques based on interaction. It is based on the approach proposed in [3], in which the behavior of vehicles at an intersection was modelled by a Dynamic Bayesian Network (DBN), and the problem of risk estimation was solved by applying a particle filter to this network. In this study, the applicability of this type of methods have been validated in situations of certain complexity, which involve more than two vehicles in scenarios such as roundabouts and intersections with several lanes. These methods imply the execution of algorithms with a high computational cost, but which can be very useful to predict the intention of other relevant participants in the vicinity of the vehicle.

5.3.1 Dynamic Bayesian Network Model

The Dynamic Bayesian Network model used in this approach is presented in Fig. 5.2, where bold arrows represent multi-vehicle dependencies, i.e. the influences of the

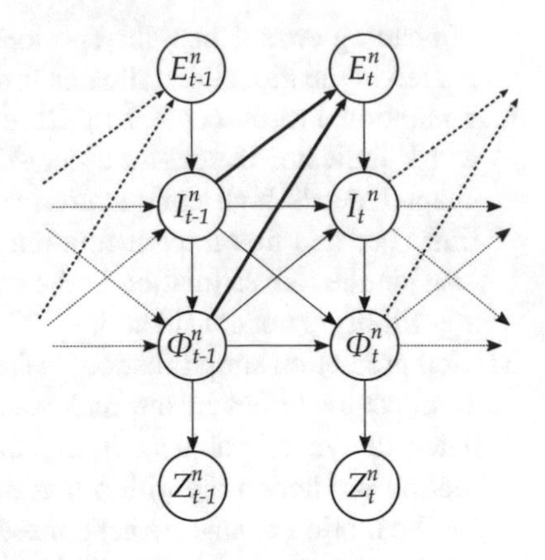

Fig. 5.2 Structure of the dynamic bayesian network

other vehicles on vehicle n. The state variables used in the model are defined in detail in the following subsection.

5.3.1.1 Definition of State Variables

As depicted in 5.2, four state variable types are used: (i) Expected manoeuvre (E^n_t), (ii) Intended manoeuvre (I^n_t), (iii) Physical State (Φ^n_t) and (iv) Measurements (Z^n_t), where the expected and intentional manoeuvres are unknown and intangible variables, unlike the physical state.

Bear in mind that the network represented in the Fig. 5.2 is instantiated for each of the vehicles. Moreover, since in a real driving scenarios the behaviour of each vehicle is conditioned by the behaviour of the rest vehicles, these networks are interrelated.

According to the general model of Fig. 5.2, the following generalized distribution is extracted:

$$P\left(\mathbf{E_{0:T}}, \mathbf{I}_{0:\mathbf{T}}, \Phi_{0:\mathbf{T}}, \mathbf{Z}_{0:\mathrm{T}}\right) = P\left(\mathbf{E_0}, \mathbf{I}_0, \Phi_0, \mathbf{Z_0}\right) \times \prod_{t=1}^{T} \times \prod_{n=1}^{N} [P(E^n_t | \mathbf{I_{t-1}} \Phi_{\mathbf{t-1}}) \times P(I^n_t | \Phi^n_{t-1} I^n_{t-1} E^n_1) \times P(\Phi^n_t | \Phi^n_{t-1} I^n_{t-1} I^n_t) \times P(Z^n_t | \Phi^n_t)] \tag{5.1}$$

The used variables are described below.

Expected Manoeuvre

The expected manoeuvre E^n_t represents the expected behaviour of vehicle n at instant t according to traffic rules. This variable depends both on the vehicle itself and on other involved vehicles, and is divided into two types: longitudinal (Es^n_t) and lateral (Ec^n_t). On the one hand, the Es^n_t longitudinal expectation is used to model the probability that the vehicle should stop at an intersection. It has two possible states, GO and STOP:

- $Es_t^n = GO$: The vehicle does not have the duty to stop. This fact can be due to an intersection that n has priority of passage, the rest of vehicles have a stop sign or yield the passage, there is no vehicle with higher priority of passage approaching the intersection, and so on.
- $Es_t^n = STOP$: The driver is expected to stop before reaching the intersection, as another vehicle approaches with higher priority or there is a stop sign.

On the other hand, the lateral expectation (Ec_t^n) models the probability that the vehicle can make a lane change without hindering traffic. It has three possibilities that can be grouped into two possible states:

- $Ec_t^n = STAY$: The vehicle should not make a lane change because traffic conditions do not allow it. This may be due to the rest of the lanes being congested, there is not enough space (gap) between adjacent vehicles, the current lane is the most suitable for the speed that the vehicle is carrying, etc.
- $Ec_t^n = CHANGE$: The vehicle can make a lane change without hindering traffic or causing an accident. In this case there are two possible sub-states, which would be the possibility to change to the right or left.

Intended Manoeuvre

The intentional manoeuvre (I_t^n) has a certain similarity to the expected manoeuvre since it models the same behaviour from the perspective of the driver himself. However, this variable models the intention of the driver instead of its expectation. As in the case of expected manoeuvre, it is divided into longitudinal and lateral. Therefore, the longitudinal intention (Is_t^n) is modelled by the probability that the driver is determined whether or not to make a stop at the next intersection. Its possible states are:

- $Is_t^n = GO$: The driver intends to continue his journey without stopping.
- $Is_t^n = STOP$: The vehicle will stop at the intersection.

The lateral intention (Ic_t^n) is the intention of the driver to change lanes or not during the next few instants. It also indicates the route the vehicle intends to follow. The lateral intention thus comprises an additional variable representing the route.

Therefore, the variables involved are the following:

- $C_t^n \in c_i, i = 1 \cdots N_c$: Indicates the route followed by the vehicle n at instant t, and is included among the possible N_c routes.
- $Ic_t^n = CHANGE$: The driver intends to change lanes.
- $Ic_t^n = STAY$: The vehicle will remain in the current lane.

Physical Vehicle State

The physical state of the vehicle n at instant t is structured in two parts: position and speed. These variables are calculated at each instant from intention. For example, if the intention in a given instant is to continue without stopping, the position in the next instant will be calculated assuming a constant speed in a straight line path.

The physical state is therefore composed of:

- P_t^n: Pose of the vehicle including position and yaw angle $(X_t^n, Y_t^n, \theta_t^n)$.
- S_t^n: Linear speed of the vehicle.

Measurements

This variable includes real measurements of the physical state of the vehicle, taken from global positioning systems, IMUs, cameras, radars or a combination of them. Unlike the previous variables, the measurements are not estimated by the probabilistic model, but are extracted from the sensors directly and compared with the variables of the estimated physical state. Thus, if the measurements and physical state coincide to a lesser or greater degree, a lesser or greater probability can be assumed that the estimates of expectation and intention are valid.

Measurements are composed of position and velocity:

- Pm_t^n: Real pose of the vehicle including position and yaw angle $(X_t^n, Y_t^n, \theta_t^n)$.
- Sm_t^n: Real linear speed of the vehicle.

5.3.2 *Particle Filter*

Particle filters provide a model for processes with unknown distribution functions, as opposed to some probabilistic methods that involve Gaussian distributions or model a priori distribution. An example of this is the Kalman filter, which optimally solves linear problems with distributions that can be approximated by a Gaussian.

For the detection of risk situations, the manoeuvre expected to be executed by the driver is compared with his intentions in a given context (e.g. intersections, roundabouts, etc). Each of the possible states of the system will be contained in what it is called particle, which will have an associated weight dependent on its proximity to reality, and which is represented as follows.

Each of the N particles will give a random value to the hidden variables, which are variables that are difficult to infer directly from externally observable variables. In our case, the hidden variables are the intention and expectation of each vehicle. From them, a prediction of the observable variables will be carried out. Once this is done for all the particles, the next step is to give a weight to each particle. This step is called updating. In this stage it is compared the extent to which the particle has succeeded in its prediction with respect to the real variable. The weight given to each particle will be greater or lesser if its prediction is more or less accurate, respectively. Thus, it is a measure of the acceptance of the hidden variables that are estimated by that particle. Particles whose prediction is reliable will have a high weight in the final prediction of the hidden variables.

Once the first measurements (Z_0) have been made, the filter work-flow starts following the sequence below:

Initialization Each particle is given a random value of both expectation and intention, and the new position is calculated from them. In order to do this, two calculation stages can be performed separately. On the one hand, the calculation of the expectation and intention to stop when the vehicle is approaching an intersection. On the other hand, the respective variables for the realization of a lane change.

Prediction & updating For each value of the vehicle pose (Pm_t^n) and speed (Sm_t^n), a normal distribution is assumed with an mean of the actual value of the position and a variance derived from the sensor used. Since real sensors are not used, the deviations have been chosen with feasible estimated values. A normal probability density function is then used to calculate the probability that an individual belongs to a given distribution. Thus, the values that come closer to the actual measurement will get a lower probability. From these probabilities the weights of the particles are updated, which will be similarly higher if the prediction resembles the measured values. This update is done by multiplying the weight of each particle at the previous instant by these probabilities. In the first simulation instant, all weights are assumed to be equal: $w_i = 1/N$, where w_i is the weight of particle i. Once the weights of all the particles have been calculated, their sum is normalised to 1.

Re-sampling The problem with the updating method used is that the weights can end up being very disparate as they evolve over the initial weight. This disparity leads to the collapse of the method, causing that particles with very small weights will become unusable in later instants. This effect is mitigated by adding a re-sampling step as proposed in [1], which is only executed when the inverse of the sum of the squared weights is below a stated value. In this way the re-sampling is carried out if there are too big weights. The aim is to choose the heaviest particles and use them to replace the others. At the end, within this group, the particles with the highest weights will be repeated more times, and those with the lowest weights will appear less often. Once this step is completed, each particle will evolve again on its own.

Result To compute the resulting intention and expectation, the weights of all the particles whose intention/expectation values are $CHANGE$ (in case of lateral estimation) or GO (in case of longitudinal estimation) are added.

These steps end with the final prediction of the hidden variables for each of the vehicles. However, the collision risk of each vehicle is still to be calculated. This risk is obtained in the last step i.e. by adding the weights of the particles whose expectation and intention differ. This occurs, for example, if the intention is to go straight at an intersection and the expectation is to stop, since this implies that, under current traffic conditions and rules, the vehicle should stop and yet intends to continue to cause an accident. The result of this calculation is the indicator of collision risk. For intersection cases, it is considered that this value should not be greater than 0.3 [2], in which case, the situation is considered as dangerous.

5.3.3 Results

In order to generate different driving scenarios, both from a geometrical point of view and from the dynamics of the agents involved in them, the SUMO open source simulator [4, 6] has been used. This simulator allows to interact with the simulation in runtime from external programs through an API called *TraCI* ("Traffic Control Interface"). Moreover, this simulator was used to extract typical speed profiles at each lane in which vehicles drive in all the tested scenarios.

To cover a significant range of cases where to assess the performance of the risk calculation, different scenarios were simulated. On the one hand, the algorithm for longitudinal risk estimation has been tested in two different scenarios: simple intersection and multi-lane intersection. On the other hand, the lateral expectation have been tested in a highway and in a multi-lane roundabout.

The procedure to validate the risk estimation algorithm is described below:

1. **Perform the simulations**. A series of driving situations has been chosen for each scenario. Simulations are carried out and data on positions, speeds and times are collected. For each simulation, it is necessary to create a set of configuration files and initial conditions.
2. **Data collection**. From the SUMO output files, a function is executed in Matlab script that passes the information to a data structure, so that it can be interpreted by the main algorithm.
3. **Execution of the risk estimation algorithm**. The algorithm is run to produce the evolution of intention, expectation and risk values over time. This step is performed several times for each simulation, since the algorithm is based on a random method and its behaviour is studied from varied results.

Once the results are obtained, they are classified by successes (the algorithm has correctly detected whether vehicles generate a hazard) or false positives (risk is detected in safe situations).

5.3.3.1 Simple Intersection

The first scenario to be tested consists of an intersection of four ways with one lane in each direction. The purpose in this test case is to evaluate the risk estimation algorithm in situations in which four vehicles interact at an intersection, yielding to vehicles with higher priority or not doing so, and thus causing or not risk situations.

The rotational symmetry (order 4) of this scenario makes possible to position the vehicle n on one of the roads and to vary the behaviour of other vehicles. The result in each case would be the same as if the position of all vehicles is rotated to the left, right or in front of them. The scenarios have been generated by using a fixed initial positioning of the vehicle n on the *CD* track and the variation in its trajectory and, on the other hand, on a variable initial positioning of the secondary vehicle that generates the risk on each of the three remaining roads. To cover most of the possible

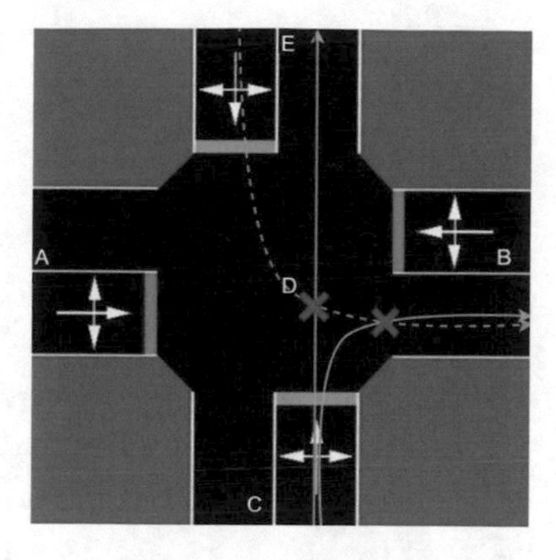

Fig. 5.3 Simple intersection scenario

combinations, 12 different scenarios have been generated. Moreover, each of these situations has been split into two possibilities: one in which the secondary vehicle generates danger in vehicle n and one in which it stops to yield the vehicle with priority. Since the risk estimation algorithm has a random component introduced by the particle filter, it is run five times for each scenario, creating a total amount of 120 tests (12 environments $\times$ 2 situations $\times$ 5 simulations).

Some of these generated situations and the trajectories of the two main vehicles are represented in Fig. 5.3, where the trajectory of vehicle n is drawn in orange.

Figure 5.4 shows the results of the algorithm in a simple intersection scenario in which one car approaches the intersection by each of its four entrances. Two of them drive on perpendicular lanes and collide at the instant $t = 22.8$ s. Figure 5.4a shows the expectation that the vehicle will stop, while Fig. 5.4b depicts the intention to continue. Thus, if both values are high, it will probably indicate that the situation is risky. In Fig. 5.4c it can be noticed that the risk increases above the stated threshold of 0.3 at instant $t = 20.7$ s (highlighted with an orange triangle in the figure), which indicates the imminent collision. The time interval between the recognition of the hazard and the instant of collision was 1.55 s. As can be seen, the value of the longitudinal expectation increases during the simulation, as both vehicles approach the intersection and the need for one of them to stop to avoid the collision becomes apparent. When the blue vehicle comes significantly closer to the junction, the algorithm is able to analyse its intention to continue, so that the risk increases until the safe limit is exceeded.

5.3.3.2 Multi-lane Intersection

This scenario extends the previous one by considering a more complex intersection that includes several lanes in each of its approaches. Moreover, the considered intersection scenario allows different possible combination of vehicles trajectories, which are shown in Fig. 5.5. Since the scenario is rotationally symmetric (order 2), the *AD*

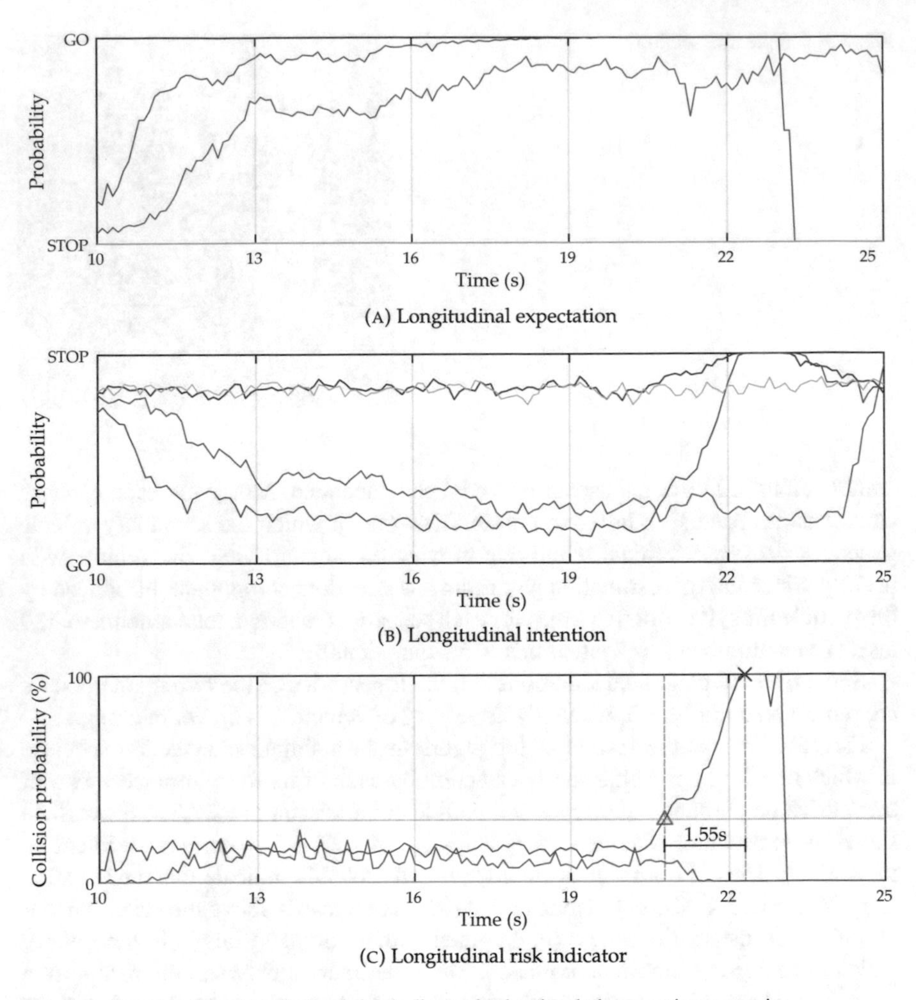

Fig. 5.4 Results of the longitudinal risk indicator in the simple intersection scenario

and *BD* streets are chosen as entrances to the priority vehicle to avoid the repetition of equivalent tests.

For each possible route of vehicle *n*, the possible situations of collision with the secondary vehicle are recreated. Figure 5.5a shows some collision possibilities, where the trajectory of the priority vehicle is represented in a continuous orange line, and the trajectory of the secondary vehicle, in a dashed blue line. As can be seen in this figure, each of the three trajectories of vehicle *n* (orange lines) collides further from the point at which the opposing vehicle begins to turn (blue line). It is to be expected that, among these three trajectories of the vehicle *n*, the further away it is from the trajectory of the oncoming vehicle, the earlier the risk situation is recognised.

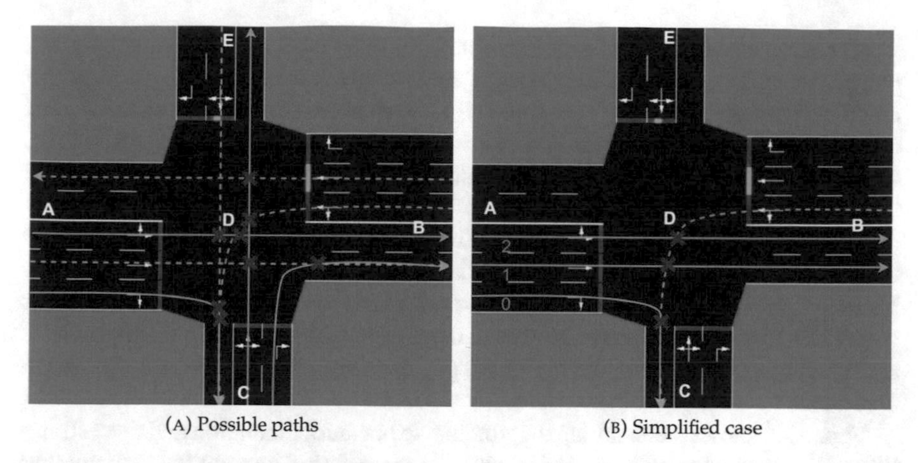

(A) Possible paths (B) Simplified case

Fig. 5.5 Multi-lane intersection scenario

Table 5.1 Time at which the collision risk is detected

Trajectory	Mean (s)	Standard deviation (s)
0	0	0
1	0.44	0.0548
2	0.82	0.0447

Table 5.2 Multi-lane intersection results

Situation	Risk detected	Risk no detected	Success rate (%)
Hazardous	85	0	100
Safe	13	87	87

Five tests were carried out for each trajectory of vehicle n with the same scenario. The resulting risk detection times are showed in Table 5.1.

As in the case of the simple intersection, in this scenario there are several situations in which the vehicle that drives in the opposite direction to the vehicle n and ends up turning left at the intersection, what is considered as non-priority path. This makes the algorithm unable to predict the risk sufficiently in advance due to ignorance of the manoeuvre intended by the other vehicle.

In addition to the scenario analysed above (depicted in Fig. 5.5b), more combination of the trajectories of both priority and secondary vehicle has been used to generate 17 different situations. Taking into account the two possible situations (danger situation for the main vehicle or not) and 5 simulations each, a total amount of 170 additional experiments were carried out. The resulting time of these cases are shown in Table 5.2, from which can be extracted that the hazardous situations are detected in all cases and some of the safe cases are detected as hazardous, causing a 13% of false positives.

Although the algorithm is able to detect all the real hazardous situations, the percentage of false positives is high, suggesting that the threshold for the risk indicator could be low for this kind of scenarios. It has been noted that the risk value of most of the false positives detections is between 0.3 and 0.4. Taking this observation into account, the false positives rate can be decreased by increasing the risk indicator threshold, although the detection time is affected. By analysing the same simulation results with a risk indicator threshold of 0.4, it can be observed a reduction of 0.1 s of the detection anticipation time while the false positives rate is reduced to 4.9%. Moreover, the tests carried out use the same particle number, regardless of the number of vehicles in the driving scene. However, it has been observed that the more vehicles are present in the scenario, the high false positives are given by the risk estimation algorithm.

Figure 5.6 shows the resulting longitudinal expectation, intention and risk estimation of the case where one car approaches the intersection thought the path number (continuous orange line in Fig. 5.5b) while another that comes from the opposite direction turns left (dashed blue line in Fig. 5.5b), causing a dangerous situation.

In comparison to the simple intersection scenario, in this case the threshold is exceeded only 0.5 s before the collision. Note that in contrast to the simple intersection, in this case the cars that generate the dangerous situation approach the intersection thought opposite directions instead of perpendicular ones.

5.3.3.3 Highway

With the highway scenario it is intended to test the risk estimation algorithm in environments where the lane change of a vehicle could cause a hazardous situation. In contrast to the previous cases, the tests carried out in this scenario focus on only three cases, shown in Fig. 5.7, since the possible traffic scenes to analyse hazardous situations in lane changing are less varied than those appearing in intersections.

In the present case, it is specially useful to order a particular vehicle (ego-vehicle) to change its lane without taking into account traffic conditions, thus obtaining dangerous situations. During the simulations, the rest of the vehicles carry out the overtaking manoeuvres simulating the human behaviour. These vehicles move at higher speeds than the vehicle that will carry out the manoeuvre in an unsafe way, so that, when changing lanes, the ego-vehicle generates danger in the vehicles that circulate at a higher speed. To illustrate the movements made by the vehicles in the simulations, Fig. 5.8 presents one of these cases step by step from top to bottom. As can be seen, this scenario consists of a one-way road with three parallel lanes.

Moreover, each scenario for each situation is simulated five times just like in previous scenarios obtaining a total a mount of 30 tests (3 environments $\times$ 2 situations $\times$ 5 simulations). In this case, the risk situations have been correctly detected by the algorithm in the 30 tests: In 15 of them the risky situation have been detected and in the other 15 tests, no dangerous situation has been noticed.

Regarding the risk prediction time, in this case all risky situations have been detected between 0.9 and 1.1 s before the lane change ends, thus providing valuable

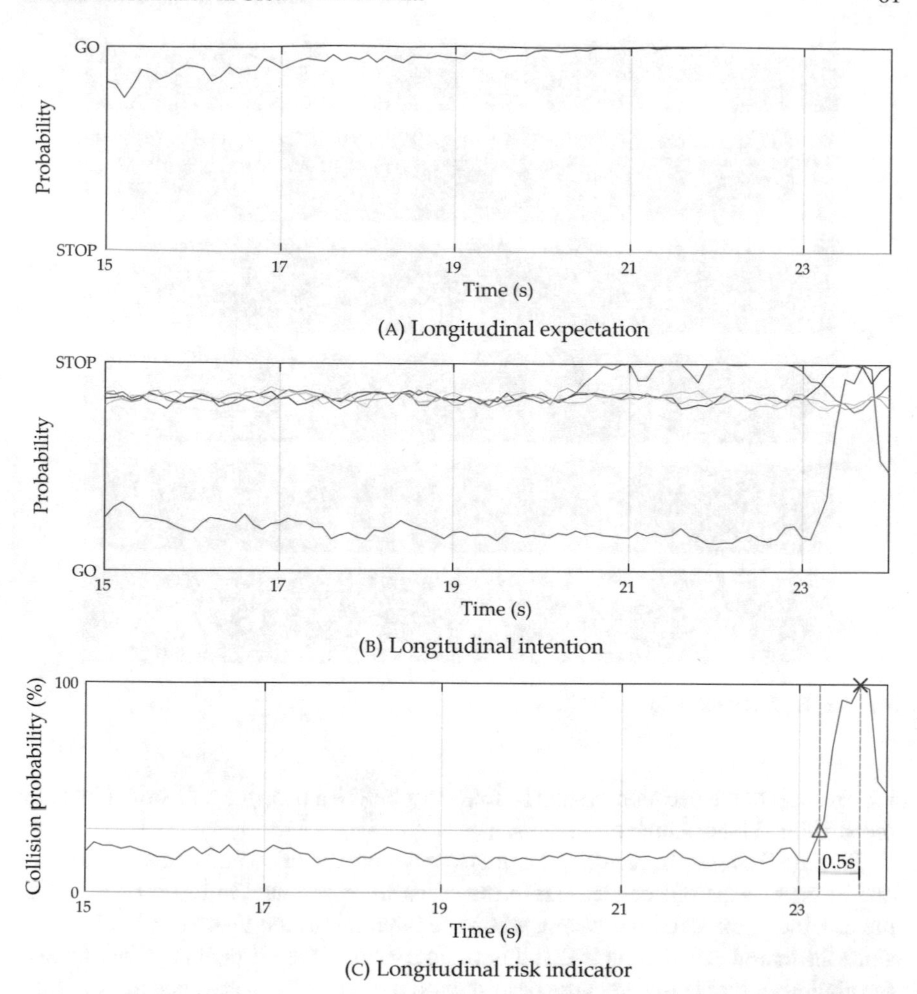

Fig. 5.6 Results of the longitudinal risk indicator in the multi-lane intersection scenario

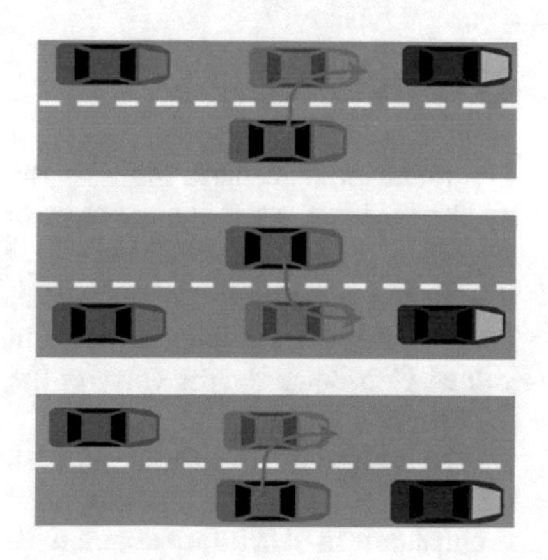

Fig. 5.7 Three considered cases in the highway scenario

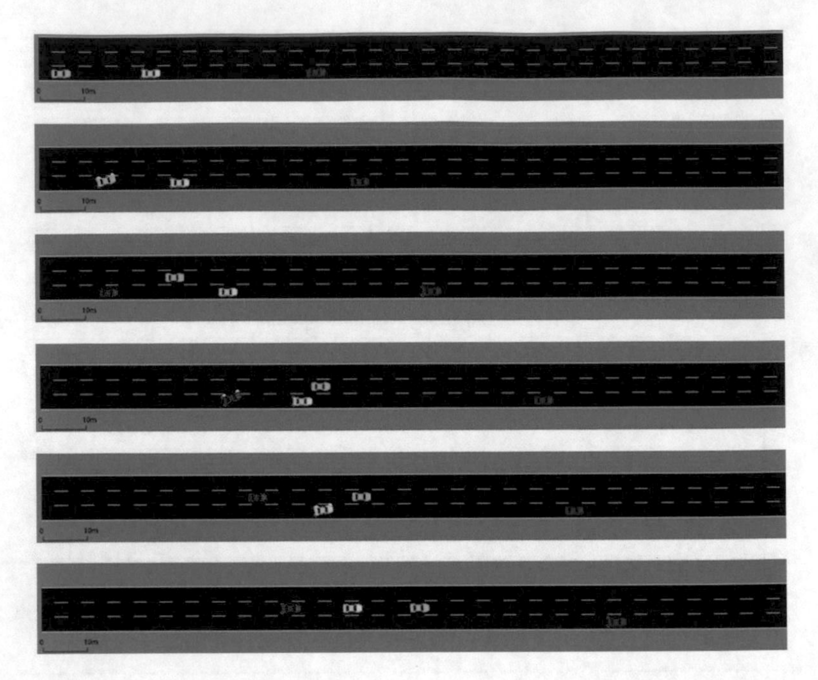

Fig. 5.8 Highway scenario

information to the decision system is able react before a possible collision. Note that the simulated lane changes last 1.5 s.

Figure 5.9 shows the resulting lateral expectation, intention and risk estimation of one of the simulations carried out in the scenario represented in Fig. 5.8. Note that the risk indicator starts increasing when the intention of the blue vehicle also does since the lateral expectation is that this vehicle stays in the original lane. In this case the algorithm is able to detect the hazardous situation 1.05 s in advance to a possible collision.

5.3.3.4 Roundabout

The roundabout scenario includes three lanes and the manoeuvres to be performed by the vehicles are the same than those of the highway scenario (see Fig. 5.10). However, a new complexity with respect to the previous scenario is introduced: the angular speed of the vehicles make difficult to differentiate whether a vehicle is manoeuvring a lane change or not, which causes the indicator to have generally larger values. Consequently, the value of the risk indicator threshold has been increased to 0.45 for this scenario.

The design of the roundabout scenario in SUMO entails some limitations of the realism of the simulated environment with a real one. As can be seen in Fig. 5.10, roundabouts in SUMO are not modelled with curvilinear path, but from concatenated

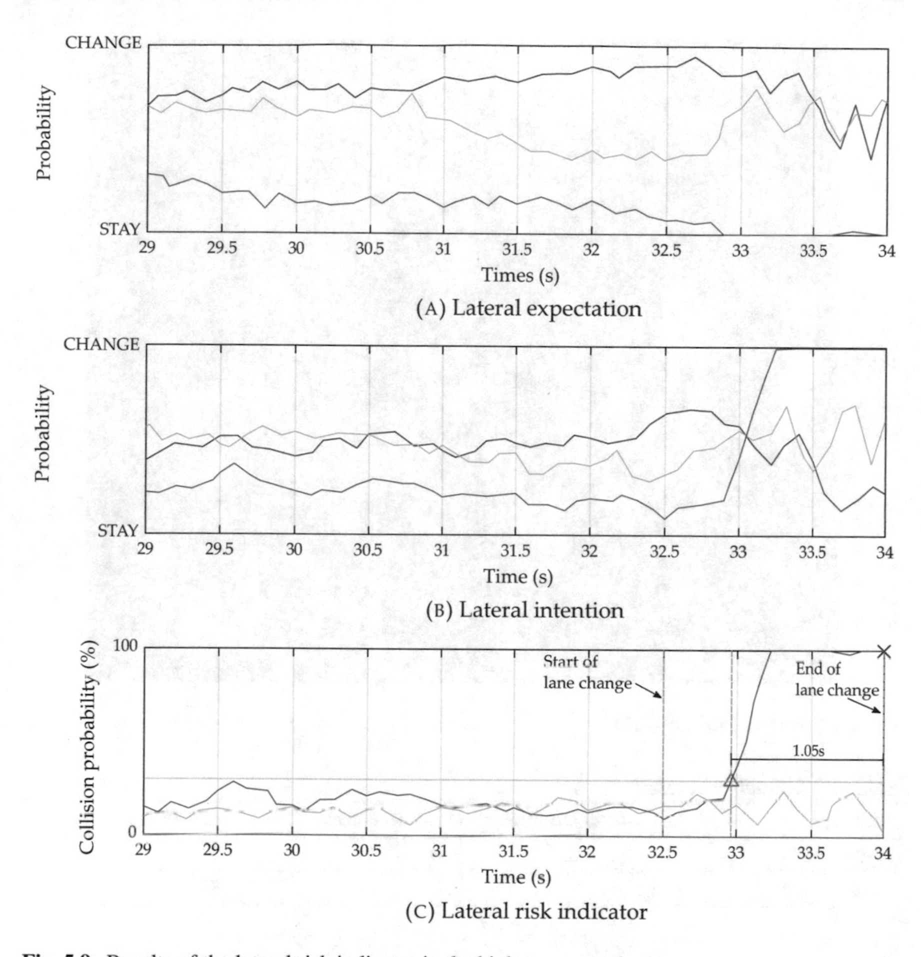

Fig. 5.9 Results of the lateral risk indicator in the highway scenario

straight segments. This way of representing a curve makes it difficult to model the position of the vehicle, since it contains completely straight sections and areas in which the turn is quite abrupt. After adjusting the modelling of the motion of the vehicle as best as possible, risk peaks continue to occur at the points where the direction of the vehicle changes, as the algorithm is not able to predict these abrupt changes.

Despite of the added complexity, the risk estimation algorithm is able to detect the risk situation properly in all the hazardous situations as shown in Table 5.3. Nevertheless, 3 of the 15 cases in safe situations are detected as risky. The high amount of false positive cases can be caused by the way in which the roundabouts are modelled in SUMO: a concatenation of straight segments. It occurs since the vehicles turn abruptly as they drive through the junction of two segments and, consequently, the risk estimation algorithm detects it as a more likely lane change.

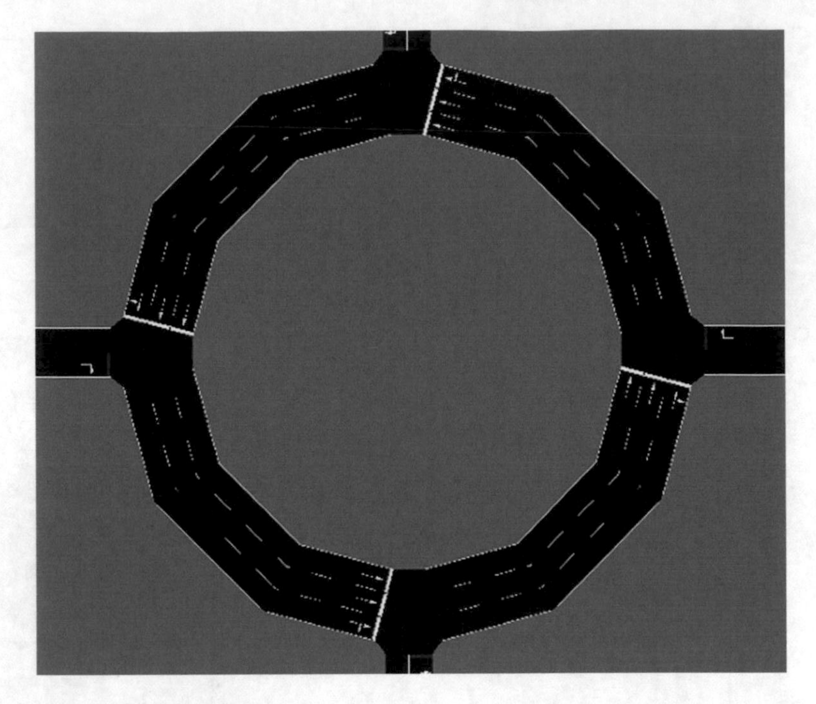

Fig. 5.10 Roundabout scenario

Table 5.3 Roundabout results

Situation	Risk detected	Risk no detected	Success rate (%)
Hazardous	15	0	100
Safe	3	12	80

Note that the simulated vehicles perform the lane change manoeuvre in 1.5 s, as in the highway scenario. The reaction times in the simulations carried out in the roundabout scenario are between 0.8 and 1.2 s.

The results of one of the simulations performed in this scenario is shown in Fig. 5.11. In this case, it can be noticed how the risk indicator exceeds the threshold risk value (0.45) 0.3 s before the possible collision.

5.3.3.5 Results Summary

As a conclusion, the results obtained in the previous subsections demonstrate the feasibility of the proposed method for all simulated situations. With the exception of the cases where it is not possible to detect the risk in advance (as in left-turnings of intersection scenarios), it predicts the risk situation 100% of the time and with acceptable margins of time. It has also been described how situations of greater com-

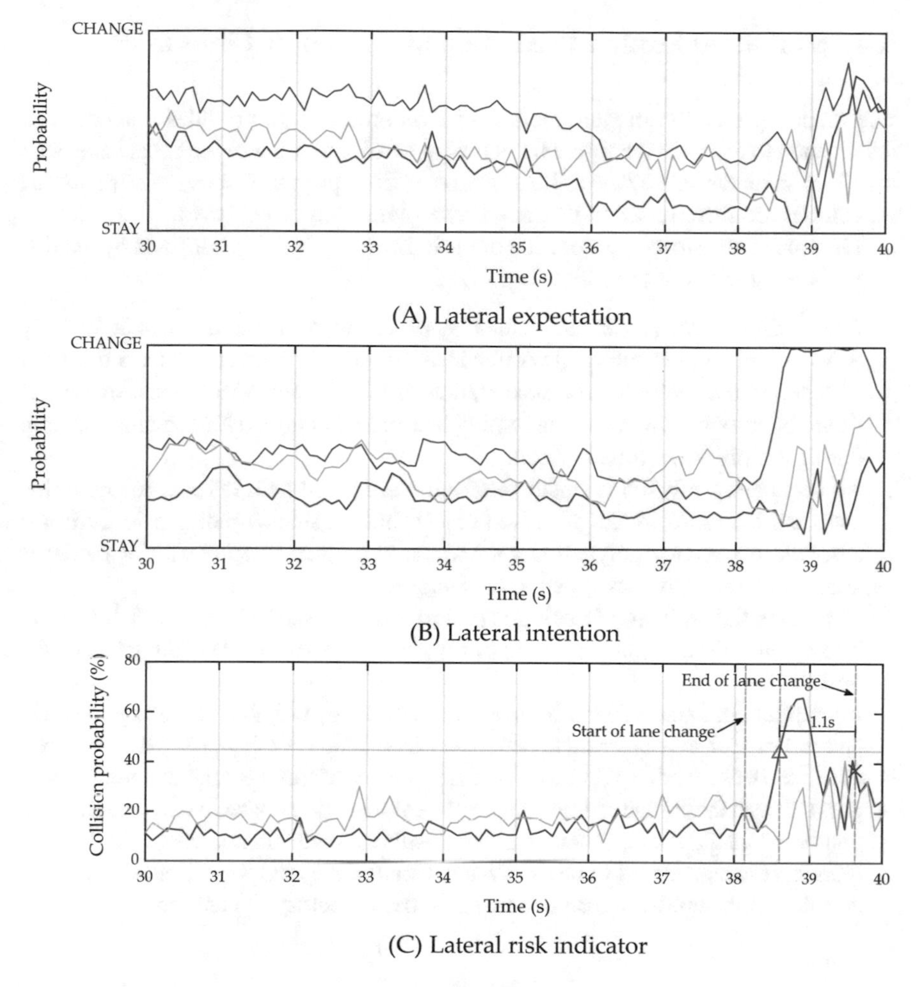

Fig. 5.11 Results of the lateral risk indicator in the highway scenario

plexity obtain worse results, since the algorithm must predict from a given number of particles a greater or lesser amount of possible situations.

In the highway and roundabout scenarios, where the lateral risk has been tested, the reaction time is below the 0.6 s. This leaves room for further improvements of the algorithm as it is expected to detect possible collisions at least 1.5 s in advance to provide the decision system with information to be able to react in time in this kind of scenarios. The number of particles then becomes the key piece to avoid the error since, the greater the number of particles, the greater the number of situations represented.

5.4 Simplified Motion Prediction for Dynamic Objects

The motion prediction module included in the architecture provides a simple and fast method to estimate the future trajectory of nearby moving objects that have been identified as hazardous by the risk estimation method proposed above. The predicted trajectories are then passed to the manoeuvre planner to be analysed.

The procedure carried out for each of the hazardous objects prioritized by the risk estimation algorithm is the following:

1. The motion of each dynamic obstacle is predicted by using a constant velocity (CV) model. As the information obtained about the perceived objects does not include angular velocity nor acceleration, this simple model is used to predict future positions of the object assuming that the velocity vector remains constant during the predicted time horizon.
2. An occupancy polygon is calculated for the predicted path in order to determine the occupied space by the perceived object that considers its dimensions. In this case, the occupancy polygon is approximated by a rectangle as a CV model is applied, so the path is assumed to be straight.
3. Then a spatial collision checking is performed as described in Sect. 6.4.2 taking into account the computed occupancy polygon instead of the bounding box of the obstacle.
4. If a spatial collision is found between the occupancy polygons of the ego-vehicle and the perceived object as depicted in the example of Fig. 5.12, the temporal collision is checked. To do that, the difference between the time the vehicle will pass through the collision point and the time the moving obstacle will pass through that point (ΔT_{coll}) is calculated. If the result is lower than a established safety margin, it is considered to be a collision. Finally, the computed time-to-collision is added to the collision data to be used in the trajectory replanning.

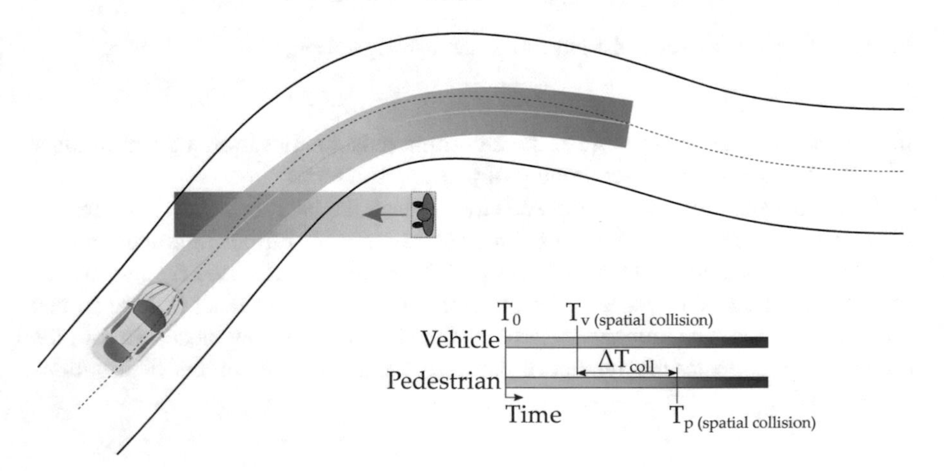

Fig. 5.12 Schematic example of motion prediction and collision detection with a dynamic object

This simplified method allows to compute future dynamic objects trajectories in a reasonable time (usually less than 1 ms) so that it can be used to check collision with the current and future ego-vehicle trajectories.

5.5 Manoeuvre Planner

In order for the vehicle to be able to react to the unexpected traffic situations, the decision-making architecture must integrate a component to monitor the perceived traffic scene. To that end, the manoeuvre planner is in charge of analysing the current state of the vehicle and the predicted behaviour of the perceived environment and consequently deciding how the vehicle should react to the current situation. To address all possible situations when analysing the predicted motion of nearby objects together with the current planned trajectory of the ego-vehicle, four different planning modes have been stipulated: (i) re-plan from current pose, (ii) extend current trajectory, (iii) avoid static obstacle and (iv) avoid dynamic obstacle. These planning modes, which correspond to those shown in Fig. 5.13, influence the trajectory generation initialization that is described in Chap. 6.

Figure 5.13 shows how these planning modes make easier the trigger of the final trajectory generation based on the current perceived situation.

In some cases, the current state of the road could impede to continue driving through the current road corridor due to road works, accidents, etc. To that end, the manoeuvre planner includes an interaction mechanism with the global planner to communicate the need of a new road corridor when needed (see figure 3.4 for more details).

As can be seen in the top of Fig. 5.13, two tasks can be performed in parallel giving as a result a new trajectory generation request or a new route request to global planner. The trajectory generation algorithm is in charge of managing simultaneous requests depending on their priority, regardless if it is already computing a new trajectory requested before.

On the one hand, based on the left branch of Fig. 5.13, the manoeuvre planner verifies firstly whether there is an existing trajectory or not and requests a new trajectory from the current vehicle pose (planning mode 0) if needed. This case only occurs in concrete situations such as the mission start. Moreover, the manoeuvre planner is continuously checking the length remaining path. If the minimum threshold established for the remaining path length (min_{pl}) is lower than this threshold, a trajectory extension is requested to the trajectory generator (planning mode 1).

On the other hand, the right branch of the flow diagram in Fig. 5.13 is devoted to analyse the perceived objects, which are classified in two groups: static and dynamic objects depending on their absolute speed. If a collision of the current trajectory of the ego-vehicle with a static object is found, information about the collision is gathered in a structure that is later passed to the trajectory generator when the planning request

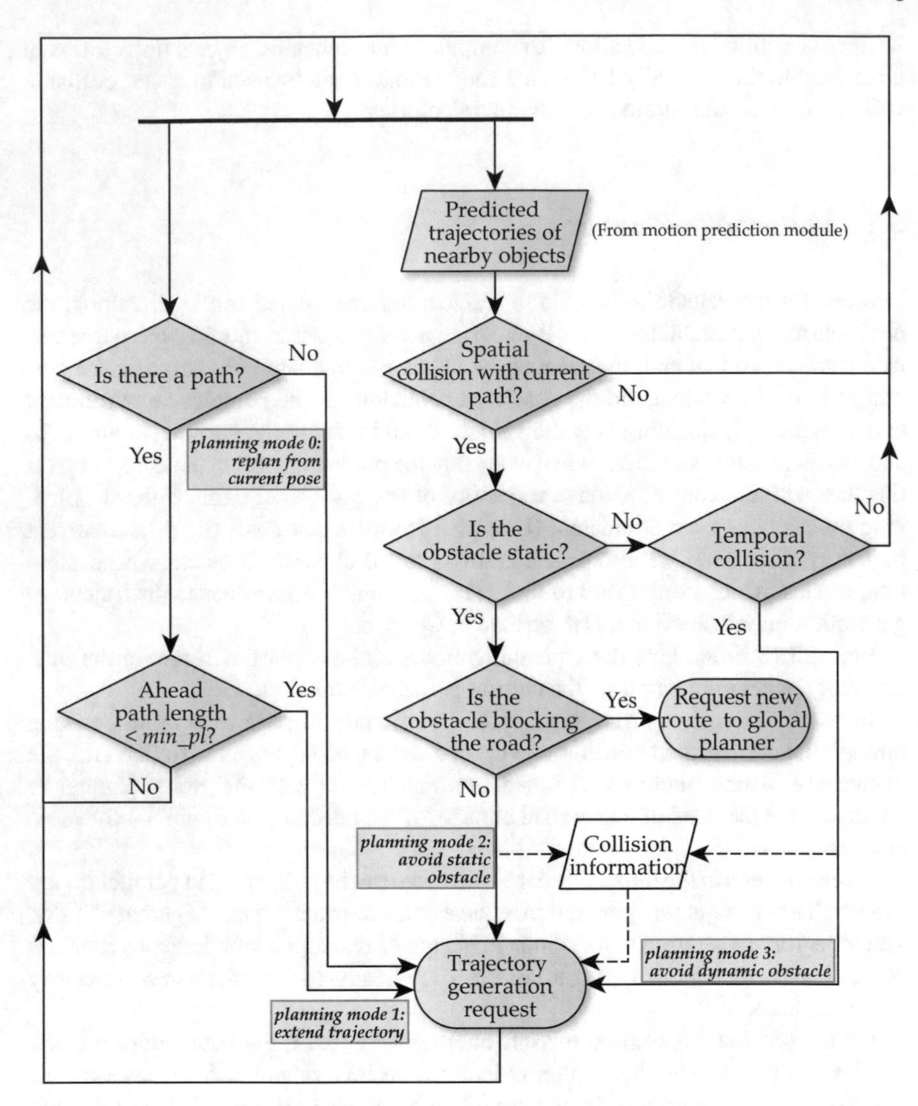

Fig. 5.13 Flow diagram of the manoeuvre planner module

is carried out. This structure contains the ID of the colliding object, time-to-collision, distance to object and the index of the current trajectory point. Based on the collision information, the trajectory generator will be able to provide a solution to the planning request.

References

1. Arulampalam MS, Maskell S, Gordon N, Clapp T (2009) A tutorial on particle filters for online nonlinear/NonGaussian Bayesian tracking. In: Bayesian Bounds for Parameter Estimation and Nonlinear Filtering/Tracking. IEEE, 2009. ISBN: 9780470544198. https://doi.org/10.1109/9780470544198.ch73. http://ieeexplore.ieee.org/search/srchabstract.jsp?arnumber=5266292
2. Lefèvre S (2012) Risk estimation at road intersections for connected vehicle safety applications. Theses. Université de Grenoble, Oct. 2012. https://tel.archives-ouvertes.fr/tel-00858906
3. Lefèvre S, Vasquez D, Laugier C (2014) A survey on motion prediction and risk assessment for intelligent vehicles. ROBOMECH J 1.1:1. ISSN: 2197-4225. https://doi.org/10.1186/s40648-014-0001-z. arXiv:1607.04788. http://www.robomechjournal.com/content/1/1/120 http://www.ncbi.nlm.nih.gov/pubmed/1569940nih.gov/articlerender.fcgi?artid=PMC364376
4. Lopez PA, Behrisch M, Bieker-Walz L, Erdmann J, Flötteröd Y-P, Hilbrich R, Lücken L, Rummel J, Wagner P, Wießner E (2018) Microscopic traffic simulation using SUMO. In: The 21st IEEE international conference on intelligent transportation systems. IEEE, 2018. https://elib.dlr.de/124092/
5. Pokorny P, Drescher J, Pitera K, Jonsson T (2017) Accidents between freight vehicles and bicycles, with a focus on urban areas. Transp Res Procedia 25:999–1007. ISSN: 23521465. https://doi.org/10.1016/j.trpro.2017.05.474. https://linkinghub.elsevier.com/retrieve/pii/S2352146517307810
6. Sánchez-Medina J, Arnay R, Artuñedo A, Campos-Cordobés S, Villagra J (2018) Simulation tools - traffic simulation. In: Intelligent vehicles. Elsevier, pp 395–436. ISBN: 9780128128008. https://doi.org/10.1016/B978-0-12-812800-8.00010-2. http://linkinghub.elsevier.com/retrieve/pii/B9780128128008000102
7. Villagra J, Perarnau M, Godoy J, Artunedo A (2018) Validación de una estrategia para la estimación del riesgo en intersecciones con vehículos conectados. In: Actas de las XXXIX Jornadas de Automática. Badajoz: Universidad de Extremadura, Sept. 2018, pp 202–209. ISBN: 978-84-09-04460-3. http://hdl.handle.net/10662/8228

Chapter 6
Optimal Trajectory Generation

6.1 Introduction

Motion planning is a core technology for autonomous driving. It must produce safe, human-like and human-aware trajectories in a wide range of driving scenarios. Whilst much progress has been attained in the perception and localization domains, digital representations of the world are still incomplete. As a result, understanding the spatio-temporal relationship between the subject vehicle and the relevant entities whilst constrained by the road network might be very difficult a challenge. Urban motion planning is significantly affected as knowledge of the environment is incomplete and the associated uncertainty is high. Most of the path planning proposals for autonomous driving assume the environment is well-known, which is rarely the case unless a specific and frequently updated high definition mapping has been carried out in the region of interest. Besides, most of the commercial digital maps for navigation use a very high-level representation, which results in far too low accuracy to obtain a good approximation of the local navigable space.

This chapter addresses the trajectory generation capability of the **local planner** section of the decision-making architecture. Bearing the above in mind, it is proposed a procedure to generate continuous curvature and minimum jerk paths from automatically generated OSM-based road corridors, so that driving trajectories are as human-like as possible. Trajectory generation component is integrated in the general decision-making architecture as depicted in Fig. 6.1.

The remainder of the chapter is organized as follows. Firstly, Sect. 6.2 states some assumptions considered in the contributions of this chapter. Section 6.3 presents an extensive analysis of interpolation curve planners, which use the current pose of the vehicle and intermediate waypoints within a short-medium time horizon. More specifically, continuous curvature Bézier-based path planners will be exhaustively compared in different driving scenarios, shedding some light into the best design choice for each particular situation and goal. Section 6.4 focuses on trajectory generation module, whose main goal is to provide a new trajectory in a short time horizon

A. Artuñedo, *Decision-making Strategies for Automated Driving in Urban Environments*, Springer Theses, https://doi.org/10.1007/978-3-030-45905-5_6

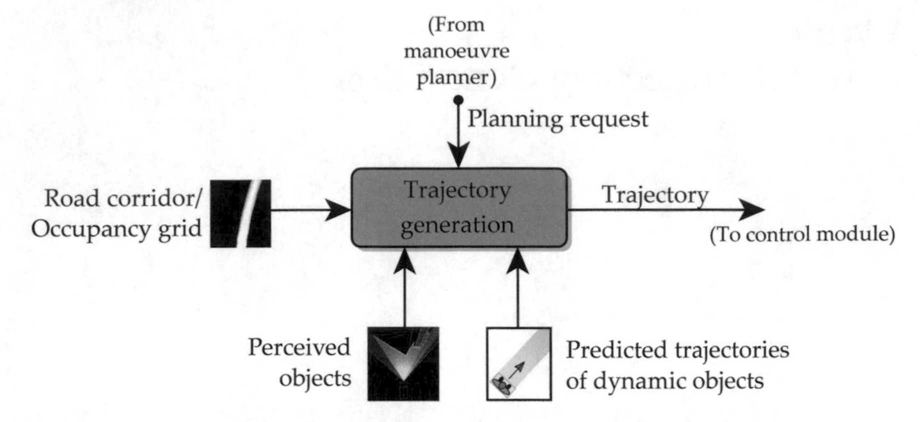

Fig. 6.1 Inputs and outputs of the trajectory generation module

when requested by the manoeuvre planner. This section introduces the path planning methods selected from the previous comparison as well as the methods used for collision checking and speed planning. Finally, Sect. 6.5 extends the capabilities of the trajectory generator presented in Sect. 6.4 to use a probabilistic occupancy grid as input for motion planning. Both modification of collision checking algorithm and path candidates evaluation are detailed in this section.

6.2 Assumptions

Some assumptions are made in the methods introduced in this chapter regarding maps, perception and localization:

Maps and road corridors The accuracy of the maps is assumed to be good enough to be used for the generation of the road corridors using the method described in Chap. 4. Taking into account the results obtained using this method, the road corridors are assumed to provide an accurate representation of the real navigable space so that they can be reliably used by motion planning algorithms.

Perception The motion planning algorithm proposed in this chapter uses the information provided by perception systems to estimate the state of nearby obstacles. The inaccuracy of this information is assumed to be low enough to use it directly by the planning strategy.

Localization The uncertainty in global localization is assumed to be negligible so it is not considered. However, Sect. 6.5 introduces a method to deal with global localization uncertainty in the motion planning process.

6.3 Optimal Path Planning

6.3.1 Problem Statement

The goal of path planning algorithms is to find a feasible path to drive from an initial point (typically the current pose of the vehicle) to a target point, while often minimizing a predefined criteria. This section focuses on path planning for autonomous driving in typical structured environments such as roads or highways, where the non-holonomic constraints of the vehicle cannot be ignored [1, 3]. The dynamic restrictions that are commonly taken into account are (i) the maximum curvature that the vehicle is able to handle and (ii) the continuity of the curvature along the planned path. It is worth to remark the importance of (ii), since discontinuities in the curvature do not allow automated vehicles to track the path. In addition to these constraints, the path is required to be comfortably driven by the vehicle i.e. the turning angle and turning speed of steering manoeuvres should not lead to strong lateral accelerations. As a result, the path planning strategy should minimize the variability of the curvature along the computed path.

Since elevation increment of the considered space for path planning is insignificant in most of the structured environments, the path planning problem is typically performed in a 2D plane. This allows to formally define the path planning problem in a general way as follows:

$$\begin{aligned} \underset{x_a}{\arg\min} \quad & J(x_a,\ D_s,\ V) \\ \text{subject to} \quad & l_b \leq x_a \leq u_b \end{aligned} \tag{6.1}$$

where:

- x_a is a vector containing all variables to be optimized. Its size and variables can vary depending on the path planning approach, as presented in following subsections.
- l_b and u_b are the lower and upper bounds of the values of x_a in order to constrain the search space of the algorithms. The bounds values depend on the scenario and the approach.
- $D_s \subset \mathbb{R}^2$ is the drivable space of scenario s.
- $V = [p_0,\ p_f,\ l_{tw},\ \kappa^v_{max}]$ includes vehicle-related information:
  - $p_0 = [x_0, y_0, \theta_0, \kappa_0]$ is the initial vehicle pose where x_0 and y_0 are the initial 2D coordinates, θ_0 the initial heading and κ_0 the initial curvature.
  - $p_f = [x_f, y_f, \theta_f, \kappa_f]$ is the final vehicle pose where x_f and y_f are the initial 2D coordinates, θ_f the final heading and κ_f the final curvature.
  - l_{tw} is the track width of the vehicle.
  - κ^v_{max} is the maximum curvature the vehicle is able to handle.

In this section, we consider different types of primitives, optimization methods and algorithms, cost functions as well as different initial and final heading and curvature configurations. These are described in detail in the following subsections.

6.3.1.1 Primitives to Compare

Interpolation curve planners use the current pose and curvature of the vehicle and some waypoints in order to obtain the final path to be followed. This path is required to have continuous curvature and has to be as much efficient as possible. To that end, different interpolation methods based on Bézier curves will be compared, always guaranteeing the continuity of the curvature (G^2 continuity) along the path. It is important to emphasise the need for G^2-continuous paths, as the position of the steering wheel is also continuous, thus improving comfort inside the vehicle.

Bézier curves present some advantages that make them suitable for path planning in autonomous driving: fast curve and curvature calculation using analytic expressions, fast collision-checking using Bézier curves properties such as convex hull property, curve-line intersection, etc. Nevertheless, diverse piecewise Bézier curves feature different stability [9], which is an important property that defines the impact of a small local change in the position of one waypoint on the whole curve shape. In general, with interpolating splines there is a trade-off between stability and higher-order [14].

Two possible variations will be explored: (i) piecewise B-splines, defined as composites of cubic Bézier curves, and (ii) quintic Bézier curves.

Cubic B-Splines Curves

They allow to generate a curve that goes through a set of given waypoints. The curve generated is a concatenation of n plane cubic Bézier sections which are defined generically as follows:

$$C_j(t) = \sum_{i=0}^{d_b} P_i^j B_{i,d_b}(t), \ t \in [0, 1], \ j = 1...n \tag{6.2}$$

being $B_{i,d_b}(t) = \binom{d_b}{i} t^i (1-t)^{d_b-1}$ the Bernstein polynomials, P_i^j the control points of Bézier section j, and d_b the degree of the Bézier curve ($d_b = 3$ in the case of cubic curves).

Continuity at joints can be guaranteed in B-splines by forcing the first and second derivatives of two contiguous Bézier sections to be equal in the joints, so that the following expression is verified:

$$2P_2^i - P_1^i = 2P_1^{i+1} - P_2^{i+1} = A_i, \ i = 1...n-1 \tag{6.3}$$

where A_i are intermediate control points. The position of these intermediate points is fixed by the $n + 1$ points to be interpolated S_i, solving the next linear equation system:

$$\begin{pmatrix} 1 & & & & & \\ 4 & 1 & & & & \\ 1 & 4 & 1 & & & \\ & 1 & 4 & 1 & & \\ & & & \ddots & 1 & \\ & & & 1 & 4 & \\ & & & & 1 \end{pmatrix} \begin{pmatrix} A_0 \\ A_1 \\ A_2 \\ A_3 \\ \vdots \\ A_{n-1} \\ A_n \end{pmatrix} = \begin{pmatrix} S_0 \\ 6S_1 - S_0 \\ 6S_2 \\ 6S_3 \\ \vdots \\ 6S_{n-1} - S_n \\ S_n \end{pmatrix} \quad (6.4)$$

Since the curvature of a cubic Bézier section is continuous, the C^2 continuity is guaranteed along the whole curve (and consequently G^2 continuity).

The curve (6.2) whose coefficients are computed solving Eqs. (6.3) and (6.4) implicitly verifies that the second derivative is zero at its initial and end points ($A_0 = S_0$ and $A_n = S_n$). However, the linear system (6.4) can be modified so as to set additional boundary conditions. In that sense, the possibilities we consider include: setting the initial and/or end tangent vector as well as setting the initial tangent and curvature vectors. These end conditions are further explained below. It is worth to mention that cubic B-splines present excellent stability as a change in one of its waypoints only affects its adjacent Bézier Sect. [8].

Initial heading setting To set the initial heading of the curve, the tangent vector at the initial point must be forced. To that end, the first derivative of (6.2) can replace the second equation of the linear system (in Eq. (6.4)):

In this case, the stability is lower than the previous case, where initial tangent is not imposed.

$$2A_0 + A_1 - 3S_0 + \vec{t}_0 \quad (6.5)$$

where $\vec{t}_0$ is the tangent vector at the initial point.

Then, the resulting linear system to solve is in Eq. (6.6).

$$\begin{pmatrix} 2 & 1 & & & \\ 1 & 4 & 1 & & \\ & 1 & 4 & 1 & \\ & & & \ddots & 1 \\ & & & 1 & 4 \\ & & & & 1 \end{pmatrix} \begin{pmatrix} A_0 \\ A_1 \\ A_2 \\ \vdots \\ A_{n-1} \\ A_n \end{pmatrix} = \begin{pmatrix} 3S_0 + \vec{t}_0 \\ 6S_1 \\ 6S_2 \\ \vdots \\ 6S_{n-1} - S_n \\ S_n \end{pmatrix} \quad (6.6)$$

Initial and final heading setting The final heading can be also forced in the same way yielding

$$\begin{pmatrix} 2 & 1 & & & \\ 1 & 4 & 1 & & \\ & 1 & 4 & 1 & \\ & & \ddots & & \\ & & 1 & 4 & 1 \\ & & & 1 & 2 \end{pmatrix} \begin{pmatrix} A_0 \\ A_1 \\ A_2 \\ \vdots \\ A_{n-1} \\ A_n \end{pmatrix} = \begin{pmatrix} 3S_0 + \vec{t}_0 \\ 6S_1 \\ 6S_2 \\ \vdots \\ S_{n-1} \\ 3S_n - \vec{t}_n \end{pmatrix} \quad (6.7)$$

where $\vec{t}_n$ is the tangent vector at the last point.

Initial heading and curvature setting In addition to the initial heading, the curvature can be also set at the initial point. In this particular case, the first and second derivatives of the B-spline Eq. (6.2) evaluated at S_0 can be written as follows:

$$A_1 - A_0 = \vec{t}_0 - \frac{1}{2}\vec{\kappa}_0 \tag{6.8}$$

$$6A_0 = 6S_0 + \vec{\kappa}_0 \tag{6.9}$$

where $\vec{\kappa}_0$ is the second derivative vector (acceleration) at the initial point of the curve. This vector can be obtained as

$$\vec{\kappa}_0 = \frac{d|\vec{t}_0|}{dt}\vec{T}_0 + \kappa_0|\vec{t}_0|^2\vec{N}_0 \tag{6.10}$$

where $\frac{d|\vec{t}_0|}{dt}$ is the rate of change of the tangent module with respect to the independent variable t of the parametric curve, and $\vec{T}_0$ and $\vec{N}_0$ are the unit tangent and normal vector at the initial point, respectively.

If constant speed (in terms of differential geometry) is considered, the tangential part of Eq. (6.10) is null, and therefore the resulting linear system (6.4) can be written as follows

$$\begin{pmatrix} -1 & 1 & & & \\ 1 & 4 & 1 & & \\ & 1 & 4 & 1 & \\ & & & \ddots & \\ & & 1 & 4 & 1 \\ 6 & & & & \end{pmatrix} \begin{pmatrix} A_0 \\ A_1 \\ A_2 \\ \vdots \\ A_{n-1} \\ A_n \end{pmatrix} = \begin{pmatrix} \vec{t}_0 - \frac{1}{2}\vec{\kappa}_0 \\ 6S_1 \\ 6S_2 \\ \vdots \\ S_{n-1} \\ 6S_0 + \vec{\kappa}_0 \end{pmatrix} \tag{6.11}$$

Note that in this case the stability is totally lost since a small change in the position of one waypoint changes the shape of the whole curve.

Quintic Bézier Splines

They have more degrees of freedom than cubic ones, which has some advantages but also drawbacks. A higher degree provides more control points, improving the controllability of the curve and consequently the controllability of the set of Bézier sections that make up the complete path. Thus, the initial and final pose (including initial and final curvature) could be imposed, making easier re-planning tasks when the vehicle is on motion and the curvature cannot suddenly change, thus maintaining G^2 continuity along the path. However, coherent values for tangent and curvature vectors at intermediate joints are also needed to place the control points of each Bézier section. As they are not known in advance, heuristic rules can be used, as proposed in [13], to estimate convenient values of first and second derivatives at intermediate waypoints ($S_i \in [S_1, S_{n-1}]$) of the quintic Bézier spline, able to guarantee the curve smoothness at joints. These estimations can be used to calculate the final curve but also can be taken as initial guesses to apply an optimization algorithm. The considered heuristics to calculate the first and second derivatives are described below.

First derivative The first derivative at waypoint S_i of the spline (tangent vector $\vec{t}_i$) is determined as follows: On the one hand, the orientation of $\vec{t}_i$ is perpendicular to the bisector of the angle formed by vector $\vec{v}_a$ and $\vec{v}_b$ where $\vec{v}_a = S_i - S_{i-1}$ and $\vec{v}_b = S_{i+1} - S_i$. On the other hand, the magnitude of $\vec{t}_i$ is set to the minimum euclidean distance between the current point (S_i) and its two neighbouring points (S_{i-1}, S_{i+1}) multiplied by a scaling factor f_t. Thus $|\vec{t}_i| = f_t\, min(|\vec{v}_a|, |\vec{v}_b|)$.

Both magnitude and orientation of $\vec{t}_i$ have a high influence on the final curve geometry. Therefore, they are variables to be optimized in some of the methods compared in this section.

Second derivative To determine the second derivative the heuristic proposed in [13] is applied. To estimate the curvature $\vec{\kappa}_i$at S_i, this approach uses two cubic Bézier sections (one from S_{i-1} to S_i and another from S_i to S_{i+1}). The second derivative is applied at S_i in both curves, using the tangent vectors at the previously mentioned points ($\vec{t}_{i-1}$, $\vec{t}_i$) and $\vec{t}_{i+1}$:

$$\vec{\kappa}_i^a = 6S_{i-1} + 2\vec{t}_{S_{i-1}} + 4\vec{t}_{S_i} - 6S_i \tag{6.12}$$

$$\vec{\kappa}_i^b = -6S_i - 4\vec{t}_{S_i} - 2\vec{t}_{S_{i+1}} + 6S_{i+1} \tag{6.13}$$

Then a weighted average is calculated with the curvature vectors of both curves at S_i, in order to obtain the estimated curvature vector $\vec{\kappa}_i^e$:

$$\vec{\kappa}_i^e = \alpha\, \vec{\kappa}_i^a + (1-\alpha)\, \vec{\kappa}_i^b \tag{6.14}$$

where $\alpha = \frac{|S_i - S_{i-1}|}{|S_i - S_{i-1}| + |S_{i+1} - S_i|}$.

The six control points of each Bézier segment can be calculated equalling first and second derivatives of the quintic Bézier equations in two consecutive sections:

$$P_0^i = S_i = P_5^{i-1} \tag{6.15}$$

$$P_1^i = S_i + \frac{1}{5}\vec{t}_i \tag{6.16}$$

$$P_2^i = \frac{1}{20}\vec{\kappa}_i + 2P_1^i - S_i \tag{6.17}$$

$$P_3^i = \frac{1}{20}\vec{\kappa}_{i+1} + 2P_4^i - S_{i+1} \tag{6.18}$$

$$P_4^i = S_{i+1} - \frac{1}{5}\vec{t}_{i+1} \tag{6.19}$$

$$P_5^i = S_{i+1} = P_0^{i+1} \tag{6.20}$$

where P_m^i ($m \in \mathbb{N} : m \in [0, 5]$ are control points of Bézier section i.

Table 6.1 Cases covered regarding the imposition of initial/final heading and curvature

	Cubic B-spline				Quintic Bézier spline
	1	2	3	4	5
Initial heading (h_0)	No	Yes	Yes	Yes	Yes
Final heading (h_f)	No	No	Yes	No	Yes
Initial curvature (κ_0)	No	No	No	Yes	Yes
Final curvature (κ_f)	No	No	No	No	Yes

Heading and Curvature at the Initial and End Points

A summary of the cases covered depending on the primitive used are shown in the table below:

6.3.1.2 Considered Path Planning Approaches

The approaches considered and compared in this section cover the most common state of the art path planning techniques that are based on Bézier curve primitives, as well as some proposed novel strategies.

They all intend to find the most suitable set of intermediate waypoints. To that end, some steps are usually carried out: firstly, the centreline of the drivable space is estimated from the drivable space boundaries. Over the centreline (i) a set of **reference points** is selected. After that, (ii) the position of the **reference points** is optimized. Some existing approaches stop at this point and compute the final path by interpolating among the optimized **reference points** by means of different curve primitives. Other approaches use the optimized **reference points** to calculate intermediate waypoints, usually called **seeding points**. These latter approaches then (iii) optimize the **seeding points** based on different methods.

Different techniques used for each of the three steps stated above are further described in following subsections.

Reference Points Selection Method (RS)

The first step is to select a set of **reference points** over the centreline. To that end, three different methods are considered:

- Equidistant points over the centreline (**E**).
- Douglas-Peucker simplification algorithm [7] (**D**). This algorithm is based on tolerance of perpendicular point-to-edge distance to extract the simplified line.
- Opheim simplification algorithm [16] (**O**). Unlike Douglas-Peucker algorithm, the search area in Opheim algorithm is constrained by both a perpendicular and a radial maximum distances.

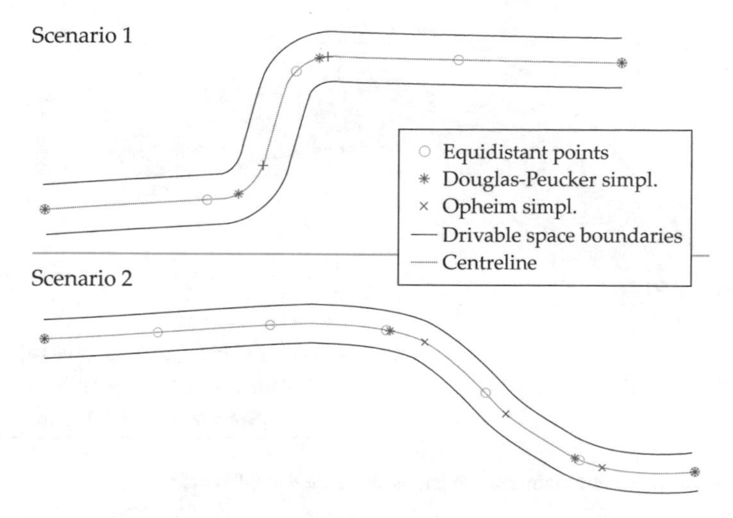

Fig. 6.2 Reference points selection: Equidistant points, Douglas-Peucker and Opheim algorithms

Table 6.2 Resultant number of selected points in both scenarios

Method	Scenario 1	Scenario 2
Equidistant points	5	7
Douglas-Peucker	4	4
Opheim	4	5

Figure 6.2 shows the results of the application of the three methods over two different scenarios. The number of points selected by each algorithms is shown in Table 6.2.

Waypoints Optimization Strategies

Both **reference points** and **seeding points** are subjected to optimization processes. Taking into account the wide range of refinement possibilities offered by the primitives considered, the approaches explained below have been addressed. For ease of reference, each method has been named with an abbreviated term.

A* search (A*) This approach is only considered for **reference points** optimization. It uses the selected **reference points** to place a set of possible waypoints over their perpendiculars to the centreline inside the drivable space boundaries. The set of possible waypoints is connected through a directed graph from the initial vehicle pose to the final intended pose, as shown in Fig. 6.3.

This graph is used to apply the A* algorithm in order to find the waypoints that minimize a cost function $f(n)$

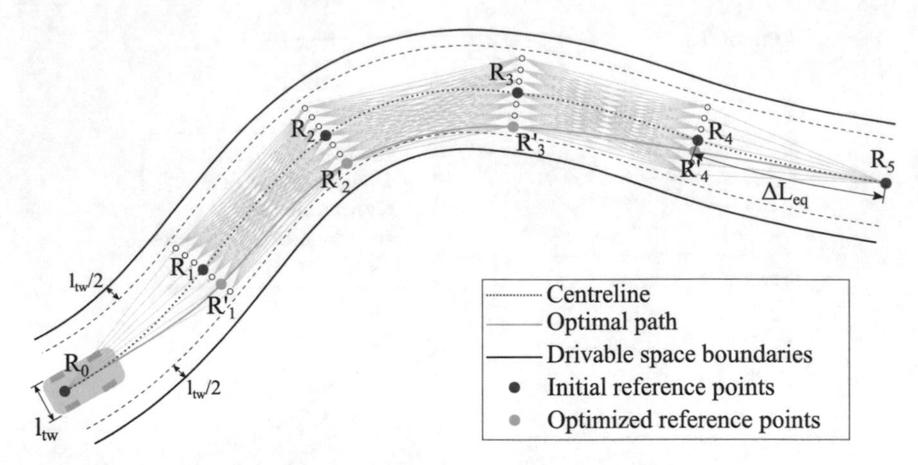

Fig. 6.3 Reference points optimization based on directed graph search

$$
\begin{aligned}
f(m) &= g(m) + h(m) \\
g(m) &= w_{koff}\ |k_e| + (1 - w_{koff})\ d_{offset} \\
h(m) &= w_d\ d_{end}
\end{aligned}
\tag{6.21}
$$

where k_e is the estimation of the curvature at the current point (m) being evaluated by using the two predecessor nodes ($m-1$ and $m-2$) to determine a circumference radius r_k ($k_e = 1/r_k$), d_{offset} is the distance from the point being evaluated (m) to the centreline, d_{end} is the distance from the point being evaluated (m) to the goal point, and weighting values w_{koff} and w_d are set to 0.5 and 0.001, respectively.

This approach is similar to [10] regarding reference trajectory planning, but in the present work curvature estimation and central offsets are considered in the cost function.

Lateral displacements (LA) The method uses lateral displacements of waypoints in order to optimize the given cost function with a specific optimization algorithm. This approach considers one continuous variable per waypoint (lateral displacements). To calculate their optimal positions R'_i, the normal vector at each waypoint R_i, $i \in [1, N]$, $N+1$, being N the number of waypoints, is used (see Fig. 6.4):

$$
R'_i = R_i + d_{lat_i} \vec{u}_{n_i}
$$

where $\vec{u}_{n_i}$ and d_{lat_i} are respectively the normal unit vectors and the lateral distance to the optimum point, computed at R_i, as shown in Fig. 6.4.

Using the notation introduced in Sect. 6.1, the optimization variables and bounds can be written for this specific case as:

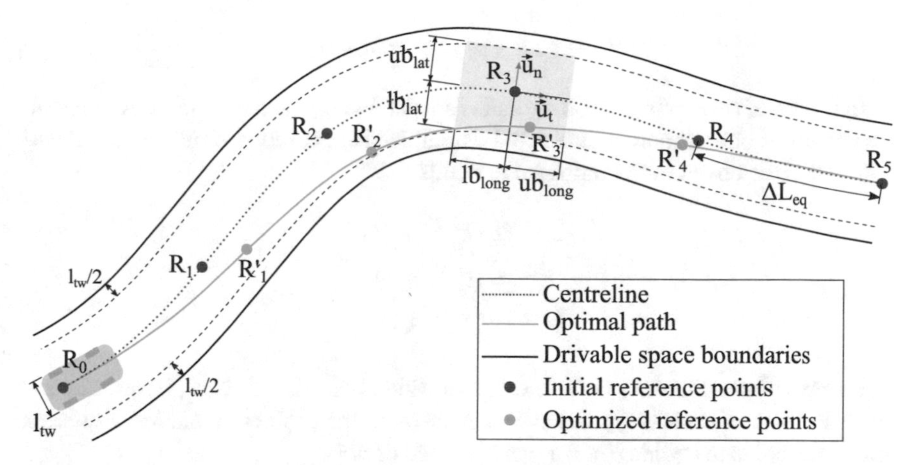

Fig. 6.4 Reference points optimization based on longitudinal and lateral movements

$$x_a^{LA} = d_{lat}$$
$$l_b^{LA} = -(\frac{l_w}{2} - \frac{l_{tw}}{2})$$
$$u_b^{LA} = (\frac{l_w}{2} - \frac{l_{tw}}{2})$$

where $d_{lat}, l_b, u_b \in \mathbb{R}^{N-1}$, being d_{lat} the set of real values containing lateral displacements of points $R_i, i \in [1, N-1]$, l_w the lane width and l_{tw} the vehicle track width.

Longitudinal displacements (LO) This method is similar to the previous one. The only difference is the use of longitudinal displacements instead of lateral ones:

$$R_i' = R_i + d_{long_i} \vec{u}_{t_i}$$

where $\vec{u_{t_i}}$ are the tangential vectors at points R_i (Fig. 6.4).

In this case, the optimization variables and bounds are:

$$x_a^{LO} = d_{long}$$
$$l_b^{LO} = -(-\Delta L/3)$$
$$u_b^{LO} = (\Delta L/3)$$

where $d_{long}, l_b, u_b \in \mathbb{R}^{N-1}$, being d_{long} the set of real values containing longitudinal displacements of points $R_n, n \in [1, N-1]$, and ΔL the distance to the closest waypoint of the two adjacent waypoints.

Lateral and longitudinal displacements (LL) This case is a combination of the two above, where lateral and longitudinal variations are considered:

$$R'_i = R_i + d_{long_i}\vec{u}_{t_i} + d_{lat_i}\vec{u}_{n_i}$$

In this case, two optimization variables are needed for each waypoint. As s result, the vectors of bounds and variables to be optimized are composed by the concatenation of the ones from both combined methods.

$$\begin{aligned} x_a^{LL} &= (x_a^{LA}, x_a^{LO}) \\ l_b^{LL} &= (l_b^{LA}, l_b^{LO}) \\ u_b^{LL} &= (u_b^{LA}, u_b^{LO}) \end{aligned}$$

Lateral displacements with selection option (LAS) In this case, besides lateral displacements of each waypoint as in **LA** method, the problem includes additional binary variables to decide if a waypoint is used or not:

$$\begin{aligned} x_a^{LAS} &= (x_a^{LA}, b_{N-1}) \\ l_b^{LAS} &= (l_b^{LA}, 0_{N-1}) \\ u_b^{LAS} &= (u_b^{LA}, 1_{N-1}) \end{aligned}$$

where $b \in 0, 1$ is a binary vector of size $N - 1$ that indicates whether a point R_i is used as waypoint to calculate the path or not, 0_{N-1} and 1_{N-1} are all-zeros and all-ones vectors of size $N - 1$, respectively.

Longitudinal displacements with selection option (LOS) This method is similar to the previous one but using longitudinal displacements instead of lateral ones:

$$\begin{aligned} x_a^{LOS} &= (x_a^{LO}, b_{N-1}) \\ l_b^{LOS} &= (l_b^{LO}, 0_{N-1}) \\ u_b^{LOS} &= (u_b^{LO}, 1_{N-1}) \end{aligned}$$

Lateral and longitudinal displacements with selection option (LLS) This approach was presented in [4]. It is a combination of the two above, where binary variables are introduced to decide if a waypoint is used or not, in addition to lateral and longitudinal displacements. Three variables per waypoint are used by the optimization algorithm in this case.

$$\begin{aligned} x_a^{LLS} &= (x_a^{LA}, x_a^{LO}, b_{N-1}) \\ l_b^{LLS} &= (l_b^{LA}, l_b^{LO}, 0_{N-1}) \\ u_b^{LLS} &= (u_b^{LA}, u_b^{LO}, 1_{N-1}) \end{aligned}$$

Tangent vector magnitude (TM) As explained in Sect. 6.3.1.1, the magnitude of tangent vector has a high impact on the curve geometry. Using this method, the magnitude of tangent vector at each waypoint is optimized within a constrained range.

$$\vec{t_i'} = f_{t_i} \vec{u}_{t_i}$$

where $\vec{u}_{t_i}$ is the tangent vector at point R_i, and f_{t_i} is the magnitude of the new tangent vector.

In this case, the path planning problem variables and bounds are:

$$\begin{aligned} x_a^{TM} &= f_t \\ l_b^{TM} &= f_{t_{min}} \cdot 1_{N-1} \\ u_b^{TM} &= f_{t_{max}} \cdot 1_{N-1} \end{aligned}$$

where $f_t \in \mathbb{R}^{N-1}$, f_t is the set of real values containing the magnitudes of tangent vectors at intermediate waypoints, $f_{t_{min}}$ and $f_{t_{max}}$ are the minimum and maximum values for the scaling factor.

Tangent vector orientation (TD) The orientation of the tangent vector highly affects the final curve too. Its value at each waypoint is optimized within a constrained range centred in the initial tangent orientation:

$$\vec{t_i'} = 1 \underline{/\theta_i} \, \vec{t_i}$$

where θ_{t_i} is the tangent vector orientation with respect to initial tangent orientation at point R_i.

In this case, the path planning problem variables and bounds are:

$$\begin{aligned} x_a^{TD} &= \theta_t \\ l_b^{TD} &= (-\Delta\theta_t \cdot 1_{N-1}) \\ u_b^{TD} &= (\Delta\theta_t \cdot 1_{N-1}) \end{aligned}$$

where $\theta_t \in \mathbb{R}^{N-1}$, θ_t is the set of real values containing the orientation of tangent vectors at intermediate waypoints and $\Delta\theta_t$ is the maximum allowed variation of tangent orientation.

Tangent vector magnitude and orientation (TT) This method is a combination of the two above, where both magnitude and orientation of the tangent vector are optimized. Just like in **LL** case, the vectors of bounds and variables to be optimized are composed by the concatenation of the ones from both combined methods.

$$\begin{aligned} x_a^{TT} &= (x_a^{TM}, x_a^{TD}) \\ l_b^{TT} &= (l_b^{TM}, l_b^{TD}) \\ u_b^{TT} &= (u_b^{TM}, u_b^{TD}) \end{aligned} \tag{6.22}$$

Curvature at joints (KJ) In this method, the curvature at intermediate waypoints is optimized. As explained in Sect. 6.3.1.1, curvature can be imposed at the joints of Bézier sections when quintic curves a re used. The approach is similar to the adopted in the **TM** method, i.e. a proportional factor (in this case f_{κ_i}) is used to modify the curvature at each intermediate waypoint as expressed in (6.22). The path planning problem variables and bounds are:

$$\begin{aligned} x_a^{KJ} &= f_\kappa \\ l_b^{KJ} &= f_{\kappa_{min}} \cdot 1_{N-1} \\ u_b^{KJ} &= f_{\kappa_{max}} \cdot 1_{N-1} \end{aligned}$$

where $f_\kappa, l_b, u_b \in \mathbb{R}^{N-1}$ is the set of real values containing the magnitude of tangent vectors at intermediate waypoints, $f_{\kappa_{min}}$ and $f_{\kappa_{max}}$ are the minimum and maximum values for the scaling factor. $f_{\kappa_{max}}$ is determined so as to ensure the maximum feasible curvature of the vehicle is not exceeded ($\kappa_i \cdot f_{\kappa_{max_i}} \leq \kappa^v_{max}$).

Tangent vector magnitude and curvature at joints (MK) A combination of **TM** and **KJ** method is also considered. In this case, the optimization problem variables and bounds are:

$$\begin{aligned} x_a^{MK} &= (x_a^{TM}, x_a^{KJ}) \\ l_b^{MK} &= (l_b^{TM}, l_b^{KJ}) \\ u_b^{MK} &= (u_b^{TM}, u_b^{KJ}) \end{aligned}$$

Tangent vector orientation and curvature at joints (DK) The last considered approach combines **TD** and **KJ** methods. the optimization problem variables and bounds are:

$$\begin{aligned} x_a^{DK} &= (x_a^{TD}, x_a^{KJ}) \\ l_b^{DK} &= (l_b^{TD}, l_b^{KJ}) \\ u_b^{DK} &= (u_b^{TD}, u_b^{KJ}) \end{aligned}$$

Note that methods **TM, TD, TT, MK** and **DK** can only be applied when using quintic Bézier splines, as cubic B-splines do not allows to impose tangent at intermediate waypoints.

As exposed at the beginning of this section, cases with two optimization stages are also considered in this study. In these cases the first stage is used to optimize the position of **reference points** from which a set of new intermediate points (**seeding points**) will be obtained. To that end, the corresponding primitive is discretized with a fixed amount of points, as depicted in Fig. 6.5.

In cases with two optimization stages only methods **A*, LA, LO, LL, LAS, LOS, LLS** are considered for **reference points** optimization, while all of them are considered for the second stage and for the cases with only one optimization process.

Optimization Algorithms Four algorithms to solve constrained non-linear multivariable optimization problems are compared: Interior-point [5], Levenberg-

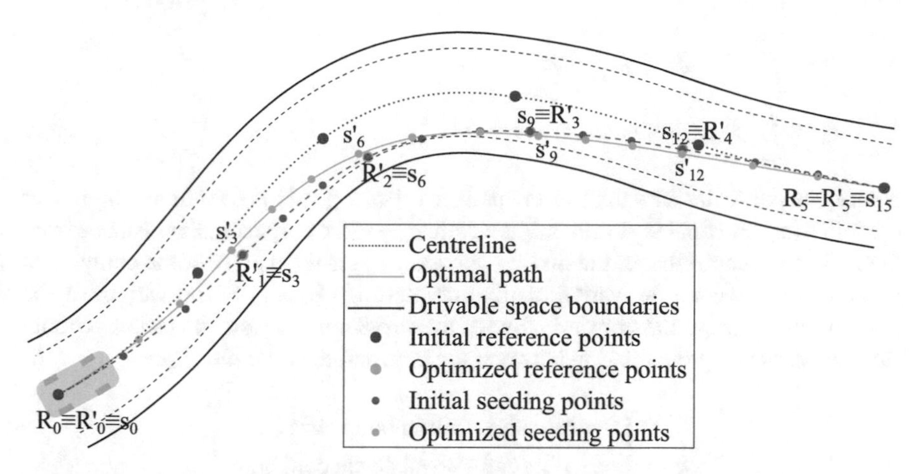

Fig. 6.5 Seeding points optimization

Marquardt [15], Simple Multi-Objective Cross-Entropy (SMOCE) [11] and Non-linear Optimization by Mesh Adaptive Direct Search (NOMAD) [6].

SMOCE is an evolutionary multi-objective optimization algorithm which presents remarkable performance in solving complex problems with many decision variables. This algorithm can be configured through four parameters: (i) epochs: number of iterations of the evolutionary process of the algorithm, (ii) working population size: amount of possible solutions evaluated by the algorithm in each epoch, (iii) elitist fraction: fraction of the working population that is selected in each epoch, and (iv) histogram intervals number: amount of intervals that are created in each dimension of the objective space. The algorithm is applied to solve single-objective problems.

NOMAD is able to solve mixed-integer non-linear programming problems (MINLP), as the ones defined in **LAS, LOS** and **LLS** methods. This algorithm is highly configurable and is designed for constrained optimization of non-linear functions.

As in path planning methods, the algorithms are referenced using shorted names: interior-point (**IP**), Levenberg-Marquardt (**LM**), NOMAD (**NM**), SMOCE (**CE**).

Cost Functions Based on the related work and the tests carried out, five cost functions are proposed based on curvature that are typically used to path planning optimization processes:

1. $J_1 = \int_{s_0}^{s_f} \dot{\kappa}(s)^2 \, ds + h$
2. $J_2 = \int_{s_0}^{s_f} \ddot{\kappa}(s)^2 \, ds + h$
3. $J_3 = \int_{s_0}^{s_f} \dot{\kappa}(s)^2 + w_{J3} \cdot \ddot{\kappa}(s)^2 \, ds + h$

4. $J_4 = \int_{s_0}^{s_f} d_{off}\, ds + h$

5. $J_5 = \int_{s_0}^{s_f} d_{off} + w_{J5} \cdot \dot{\kappa}(s)^2\, ds + h$

where κ is the scalar curvature of the path, $s \in [s_0, s_f] \in \mathbb{R}$ is the curve length over the initial (s_0) and final (s_f) values of the path, d_{off} is the perpendicular distance from the path to the centreline of the driving corridor, w_{J3} is the weight of the component related to the second derivative of the curvature in J_3, w_{J5} is the weight of the component related to the first derivative of the curvature in J_3, and h is a non-smooth function describing the relation between the path and the drivable space.

$$h = \begin{cases} 0 & \text{if path is within boundaries} \\ \infty & \text{if boundaries/obstacle collision} \\ \infty & \text{if } \kappa^p_{max} \geq \kappa^v_{max} \end{cases}$$

The value of κ is computed using the generic curvature equation of a given planar curve $c(t) = [x(t), y(t)]$:

$$\kappa = \frac{\dot{x}(t)\ddot{y}(t) - \dot{y}(t)\ddot{x}(t)}{\left(\dot{x}(t)^2 + \dot{y}(t)^2\right)^{\frac{3}{2}}} \tag{6.23}$$

where x and y are the parametric equations of the Bézier curves for both dimensions of curve c.

Note that the value of the objective function is infinity if: (i) any part of the final path is outside the boundaries, (ii) the final path collides with obstacles, or (iii) the maximum curvature along the path κ^p_{max} is greater than the maximum curvature the vehicle can handle κ^v_{max}.

6.3.2 Comparison Framework Description

Due to the large number of possibilities when composing a followable path, one of the main issues to compare different approaches to solve the same path-planning problem is to extract objective measurements that characterize them. Furthermore, the performance can be evaluated both in terms of the tracking quality of the resulting path, which can be hard to objectively define, or in run-time terms.

In order to compare all path planning methods, a common framework is specified. On the one hand, some key performance indicators (KPI) are defined with the aim of reflecting the quality of the approach employed, regardless of the type of primitive used, and not only in terms of functional performance but also of the execution time of the approaches.

6.3.2.1 Definition of KPIs

The definition of representative KPIs for benchmarking different approaches is not a trivial task. There is no a clear objective way of assessing a path planner. In fact, it is common to find a useful KPI for a small set of cases in specific scenarios that is useless or gives a wrong indication in other cases.

Besides computational cost, safety and comfort will be considered with different metrics related to the curvature and its variation. In addition to that, tunability and stability of the resulting path will be also taken into account through the offset to the centreline. Based on the tests carried out when developing this work, the most suitable KPIs contemplating all these aspects are the following:

1. Execution time: $K_t = t_{exc}$
2. Maximum curvature: $K_{\kappa_{max}} = \kappa_{max}$
3. Normalized accumulated curvature along the path:

$$K_{\kappa 0} = \frac{1}{L_p} \int_{s_0}^{s_f} \kappa(s)^2 \, ds \tag{6.24}$$

4. Normalized accumulated first derivative of the curvature along the path:

$$K_{\kappa 1} = \frac{1}{L_p} \int_{s_0}^{s_f} \dot{\kappa}(s)^2 \, ds \tag{6.25}$$

5. Normalized accumulated second derivative of the curvature along the path:

$$K_{\kappa 2} = \frac{1}{L_p} \int_{s_0}^{s_f} \ddot{\kappa}(s)^2 \, ds \tag{6.26}$$

6. Centreline offset along the path:

$$K_{cl} = \frac{1}{L_p} \int_{s_0}^{s_f} d_{off} \, ds \tag{6.27}$$

where $L_p = \sum_{i=1}^{N} \|p_i - p_{i-1}\|$ is the length of the path, p_i $(i \in \mathbb{N} : i \in [1, N])$ is the point i of the path, κ is the curvature of the path, κ_{max} is the maximum value of κ along the path, s $(s \in \mathbb{R} : s \in [s_0, s_f])$ is the curve length over the path, and t_{exc} is the total execution time.

6.3.2.2 Path Planning Problem Specification

Each path planning strategy is defined as a set of parameters related to its main characteristics. As previously described, each case is characterized by: (i) the **reference points** selection method (over a given centreline), (ii) the type of primitive curve,

(iii) the method of the first optimization process, (iv) method of second optimization process, and (v) the setting of initial and/or final heading and/or curvature as described in Sect. 6.3.1.

For the sake of clarity, each test carried out in this comparison is identified with a unique text string composed of a set of sub-IDs separated by colons. The sub-IDs are taken from the abbreviated terms specified in Sect. 6.3.1.2 for each path planning step. The complete test ID is composed as follows:

$$\text{ID} = \text{S:RS:P:O1:O2:H:K}$$

where

- **S** is the scenario number.
- **RS** is **reference points** selection method: **E**, **D** or **O**.
- **P** is the primitive type: **3** if cubic B-spline or **5** if quintic Bézier splines is used.
- **Reference points** optimization process (**O1**): It is composed of the optimization method, the optimization algorithm and the cost function. If it is set to 0, no **reference points** optimization process is carried out. **O1** is therefore defined by the concatenation of:
  - Optimization method: **A*, LA, LO, LL, LAS, LOS, LLS.**
  - Optimization algorithm: **IP, LM, NM, CE.**
  - Optimization cost function: **J1, J2, J3, J4, J5.**
- **Seeding points** optimization process (**O2**): It is composed of the optimization method, the optimization algorithm and the cost function. If it is set to 0, no **seeding points** optimization process is carried out. **O2** is therefore defined by the concatenation of:
  - Optimization method: **LA, LO, LL, LAS, LOS, LLS, TM, TD, TT, KJ, MK, DK.**
  - Optimization algorithm: **IP, LM, NM, CE.**
  - Optimization cost function: **J1, J2, J3, J4, J5.**
- **H** indicates if the initial and final heading is imposed. It is defined using two binary digits: the first one refers to the initial heading and the second one to the final heading. The considered possibilities are: **00, 10, 11**, as specified in Table 6.1 ($h_0, h_f = \{0, 1\}$).
- **K** indicates if the initial and final curvature is imposed in the same way that **H**. The considered possibilities are: **00, 10, 11**, as specified in Table 6.1 ($\kappa_0, \kappa_f = \{0, 1\}$).

One example of ID could be: 1:D:3:0:LA-IP-J1:11:00. This case addresses the scenario 1, the **reference points** are selected with the Douglas-Peucker algorithm, cubic B-splines are used as primitive curve, there is no **reference points** optimization (**O1** = 0, so the **seeding points** = **reference points**), the **seeding points** are optimized using **LA** method (lateral position optimization) with the interior-point algorithm (**IP**) and minimizing **J1** cost function. The initial and final orientation are set and the initial and final curvatures are not set.

6.3.3 Experiments and Results

6.3.3.1 Tests Cases Setup

In order to compare the performance of all the strategies presented in Sect. 6.3.1.2, all feasible combinations among the considered methods, primitives, optimization algorithms, etc. are tested in two different scenarios. As some combinations are not possible, the tests cases that include the configurations listed below are excluded:

- Impose initial curvature without imposing initial tangent.
- Impose final curvature without imposing final tangent.
- Impose final tangent without imposing initial tangent.
- Use cubic B-splines and
  - Impose final curvature.
  - Impose final tangent when initial tangent and curvature are already imposed.
  - The second optimization method is one of these: **TM, TD, TT, KJ, MK, DK** (They only apply to quintic Bézier spline cases).
- Use quintic Bézier splines and initial and/or final tangent and/or curvature are not imposed.
- There is only one optimization stage and it is defined as the first one instead of second one (**seeding points**).
- Tests cases in which there are two optimization stages and **CE** optimization algorithm is used in the second one. These cases are excluded because **CE** does not allow to set as the initial point the output of the first optimization stage.

Once the above test cases were excluded, 90417 tests per scenario were executed (180834 for both scenarios). All path planning approaches were implemented in Matlab and the experiments were executed on an Intel Core i7-3770 3.4 GHz machine with 8 GB RAM.

In order to test the path planning methods in realistic scenarios, both driving environments were extracted from real roads, which are shown in Fig. 6.2. **Scenario 1** contains two tight curves and a centreline length of 40.02 m, while **Scenario 2** comprises a roundabout entrance, with a centreline length of 54.79 m. The lane width in both scenarios is 3 m and a vehicle track width of 1.71 m (l_{tw}) is considered.

The motivation to choose scenarios with these values of centreline length is mainly due to the trade-off between computational cost and anticipation capabilities, as larger paths may involve more intermediate waypoints and therefore longer computation time. Based on the tests carried out in this study, paths with a length around 50 m are computed in a reasonable amount of time as shown in Sect. 6.3.3.2.

To determine the acceptable values for the large amount of parameters of different methods, algorithms, etc., specific tests were carried out. Regarding reference points selection, the length to obtain equidistant point was set to 7.5 m. In Douglas-Peucker algorithm, ϵ_{simp} was set to 1 m. Opheim algorithm was parametrized with minimum and maximum tolerance of 1.8 m and 30 m, respectively. The maximum function evaluation of optimization algorithms were limited in order to avoid large execution times in cases were the algorithms cannot find a solution. In the case of NOMAD algorithm the maximum optimization time was set to 20 s. SMOCE algorithm parameters was set to 50 epochs, working population size of 100 and 0.1 as elitist fraction. The epoch number and the working population size have a remarkable influence on the execution time.

In order to set an appropriate value to the parameters of cost functions defined in Sect. 6.3.1.2, different experimental test were carried out. Based on some tests carried out

Finally, the values for the parameters of cost functions **J3** and **J5**, as defined in Sect. 6.3.1.2, were empirically determined based on extensive tests carried out. Thus, w_{J3} and w_{J5} were set to 60 and 100, respectively.

6.3.3.2 Results and Discussion

Figure 6.6 represents the tests cases distribution of both scenarios using the percentiles against the value of each KPI plotted in logarithmic scales. The distribution of the tests regarding K_t and K_{cl} is similar in both scenarios. However, the values of $K_{\kappa 0}$, $K_{\kappa 1}$, $K_{\kappa 2}$ and $K_{\kappa max}$ in **Scenario 2** are generally below those in **Scenario 1**, probably due to its tighter curves. On another issue, the noticeable change in the distribution at $K_t = 20\,s$ is caused by the maximum imposed optimization time when using NOMAD algorithm.

In order to extract relevant results from all tested cases, a set of test cases of each scenario is filtered separately, based on the values of their KPIs. The thresholds used to select the minimum acceptable results with respect to the KPIs are listed below:

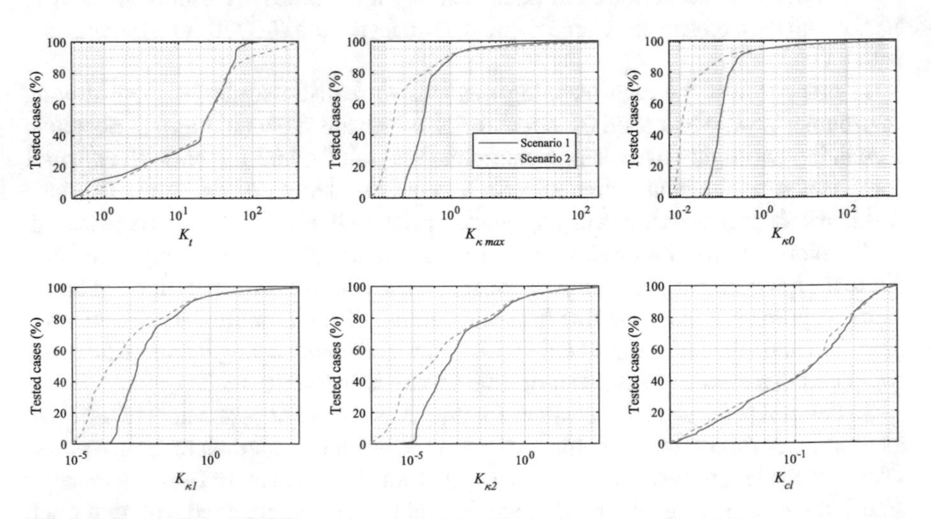

Fig. 6.6 Percentiles of all KPIs against their values for both scenarios (1 and 2)

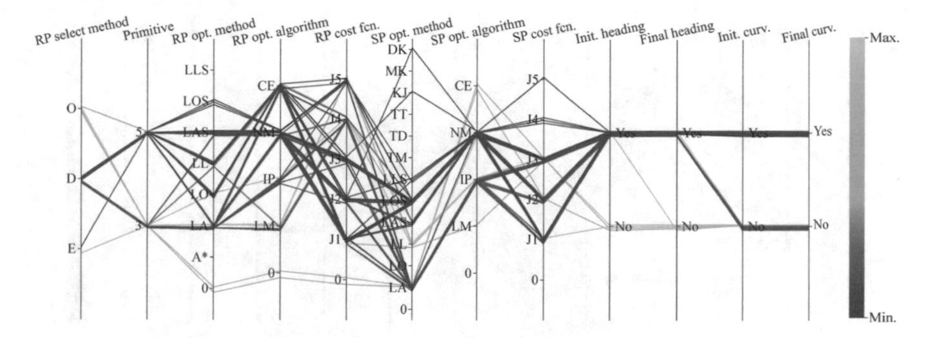

Fig. 6.7 Results of filtered tests cases with thresholds on K_t, $K_{\kappa max}$ and path length, and percentile 50th on K_{cl}, $K_{\kappa 0}$, $K_{\kappa 1}$ and $K_{\kappa 2}$ in Scenario 1. Colour is based on KPI K_t, ranging from 0 s to 50 s

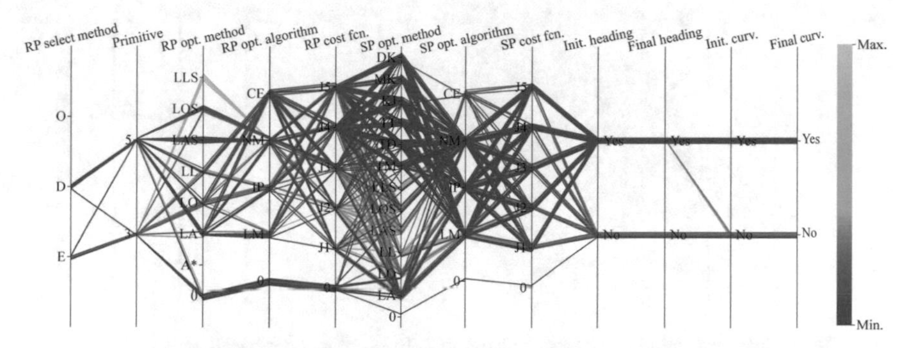

Fig. 6.8 Results of filtered test cases in Scenario 2. Colour is based on KPI K_t, ranging from 0 s to 50 s

- $K_t \leq 50$ s
- $K_{\kappa max} \leq 0.4$ m^{-1}
- $K_{\kappa 0}$, $K_{\kappa 1}$, $K_{\kappa 2} \leq 3$
- The path length is also constrained such that it does not differs more than a 5% with respect to the centreline length of the scenario (L_{cl}): $|L_p - L_{cl}| \leq L_{cl} \cdot 0.05$

The wide range of combinations of tests configurations results in high dimensional data, and therefore specific graphic representation tools are needed. By using parallel coordinates plots, Figs. 6.7 and 6.8 allow to represent the resulting KPI values in terms of its configuration values, as specified in the test ID (see Sect. 6.3.2.2), for **Scenario 1** and **Scenario 2**, respectively. The colour of each line represent the value of a particular KPI (in these two particular cases K_t). In addition to the filtering above described, the 50th percentiles of K_{cl}, $K_{\kappa 0}$, $K_{\kappa 1}$ and $K_{\kappa 2}$ were used to select the cases plotted in Figs. 6.7 and 6.8.

As can be noticed, after applying the filters there are significantly more valid tests cases in **Scenario 2** than in **Scenario 1** (1117 and 77, respectively). Furthermore, it is also remarkable that almost all cases in **Scenario 1** needed two optimization stages. Just a few of them used only one stage but with a bad performance regarding

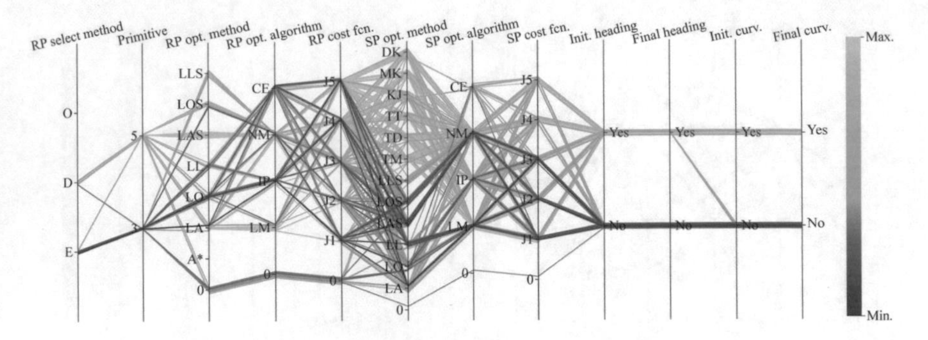

Fig. 6.9 Results of filtered test cases in Scenario 2. Line colours are based on KPI $K_{\kappa 0}$

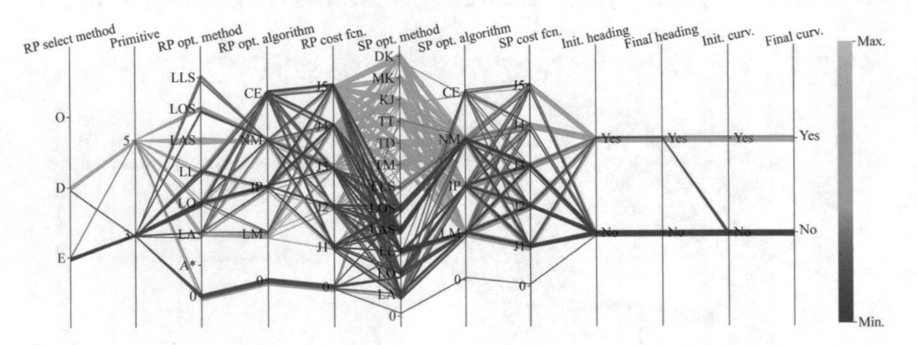

Fig. 6.10 Results of filtered test cases in Scenario 2. Line colours are based on KPI $K_{\kappa 1}$

the execution time. In contrast, although most of the cases used two optimization stages, the ones with only one optimization stage have a good timing performance in **Scenario 2**. These are signs of the greater complexity of the path planning problem in **Scenario 1** compared to **Scenario 2**.

Comparing both scenarios regarding the execution time, it is generally lower when using this Douglas-Peucker algorithm as **reference points** selection method (as shown in Table 6.2). This is non surprising, as a lower number of optimization variables is used in those cases.

Since the richness of information in **Scenario 2** is higher, i.e. there is a greater number of valid tests cases compared to **Scenario 1**, the subsequent discussion is focused on this particular information. The colour of the lines in Figs. 6.9, 6.10, 6.11 and 6.12 show the values of KPIs $K_{\kappa 0}$, $K_{\kappa 1}$, $K_{\kappa 2}$ and K_{cl}, respectively. It can be appreciated that the cases using **LA, LO, LL, LAS, LOS, LLS** methods as **seeding points** optimization present a better performance regarding $K_{\kappa 0}$, $K_{\kappa 1}$, $K_{\kappa 2}$ and K_{cl}. Contrastingly, a worse performance regarding K_t is observed when these methods are used (Fig. 6.8), where methods **TM, TD, TT, KJ, MK, DK** (only applied when using quintic Bézier splines) present better results.

The resulting paths of two selected tests cases are shown in Figs. 6.13 and 6.14. The high smoothness of optimized paths is observed in both cases. To demonstrate

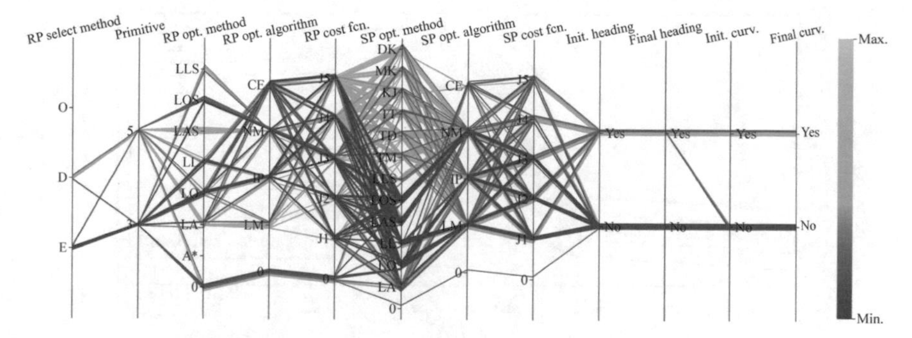

Fig. 6.11 Results of filtered test cases in Scenario 2. Line colours are based on KPI $K_{\kappa 2}$

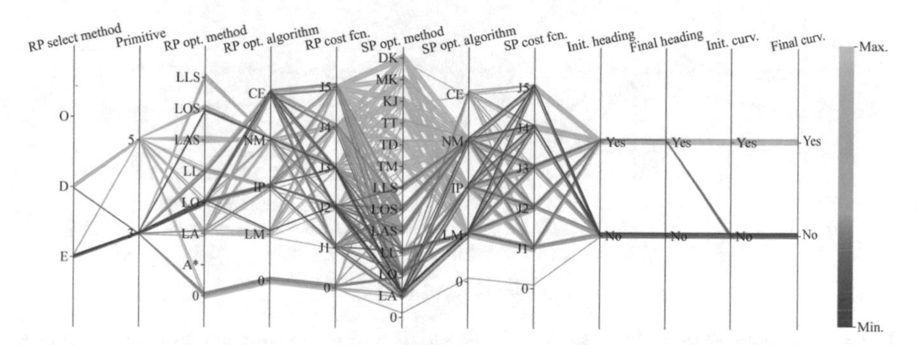

Fig. 6.12 Results of filtered test cases in Scenario 2. Line colours are based on KPI K_{cl}

that, as curve smoothness is hard to see on a 2D plane curve directly, the resultant path curvature is shown together with its first derivative at the bottom of both figures. As can be noticed, the curvature remains continuous along the path in both figures. These are two random test cases, but an insight into the best configurations is shown below.

Comparing Best Cases in Both Scenarios Delving deeper into the results, several cases with the minimum value in all KPIs are selected. The resulting KPIs are shown in Table 6.3. Moreover, the normalized KPIs of both scenarios are represented in radar charts of Fig. 6.15. Note that the normalization was done to achieve scenario-independent results, as **Scenario 1** has tighter curves, and curvature-related KPIs values are therefore higher than in **Scenario 2**.

The first thing that can be noticed from the Table 6.3 is that in both scenarios the **reference points** selection method and primitive used in all selected cases are Douglas-Peucker and quintic Bézier splines, respectively. Furthermore, it is also remarkable that SMOCE and NOMAD algorithms are also used in all scenarios for the first and the second optimization stage, respectively.

Regarding (K_t) it can be seen that test cases with ID = 1:D:5:LA-CE-J2:LOS-NM-J3:11:11 and 2:D:5:LO-CE-J3:TD-NM-J5:11:11 are those which take longer to perform the optimization of the selected tests in each scenario. However, in these

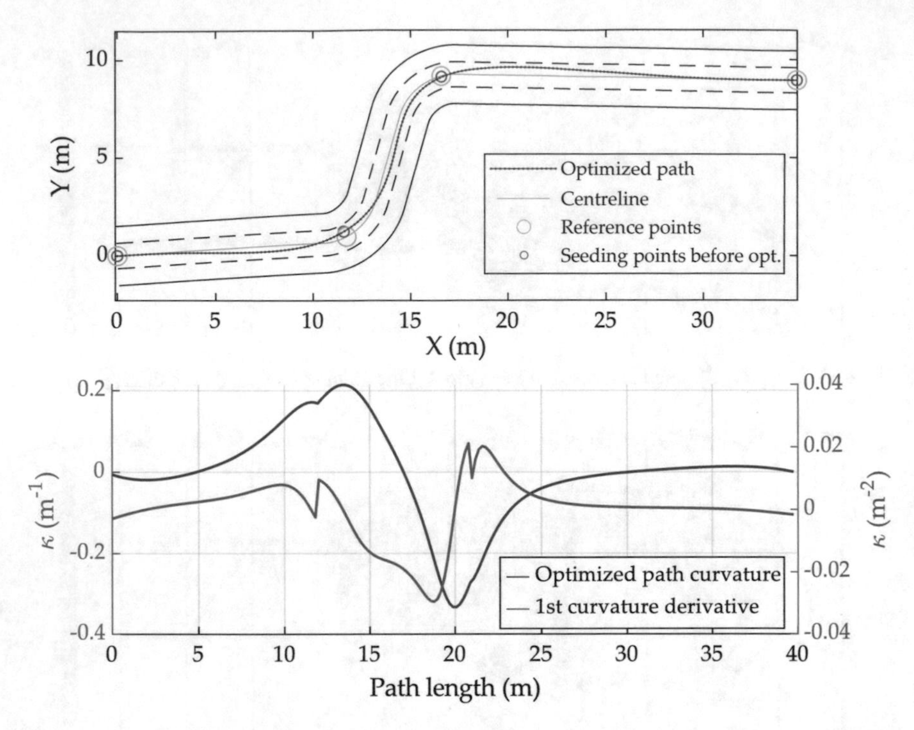

Fig. 6.13 Results of test 1:D:5:LA-CE-J1:LOS-NM-J1:11:11. Scenario and final optimized path (top) and curvature of the final path and its first derivative (bottom)

2 specific configurations almost all values of the KPIs related to the quality of the path are much lower than in the other cases. Focusing on **Scenario 1**, it is also remarkable the influence of a greater execution time in the higher quality of the final path. Regarding **Scenario 2** it can be observed that the cases 2:D:5:LA-CE-J2:TD-NM-J3:11:11 and 2:D:5:LA-CE-J2:KJ-NM-J1:11:11 present a high value in almost all KPIs, i.e. a higher execution time did not lead to better results in the quality of the final path.

Comparing Best Cases with One and Two Optimization Stages in both Scenarios In order to analyze the impact of each approach, the best results with one and two optimization stages are selected from both scenarios separately. Figure 6.16 shows the cases with one and two optimization stages in **Scenario 1**, while Fig. 6.17 depict those in **Scenario 2**.

The values of the KPIs are normalized with respect to each scenario separately, as in the previous analysis.

In this case, the tests shown in Fig. 6.17B are the same that were selected in Fig. 6.15B, where the best results present two optimization stages. However, the KPIs are normalized together with the selected tests with one optimization stage for this scenario. As a result, the represented values are different between both figures.

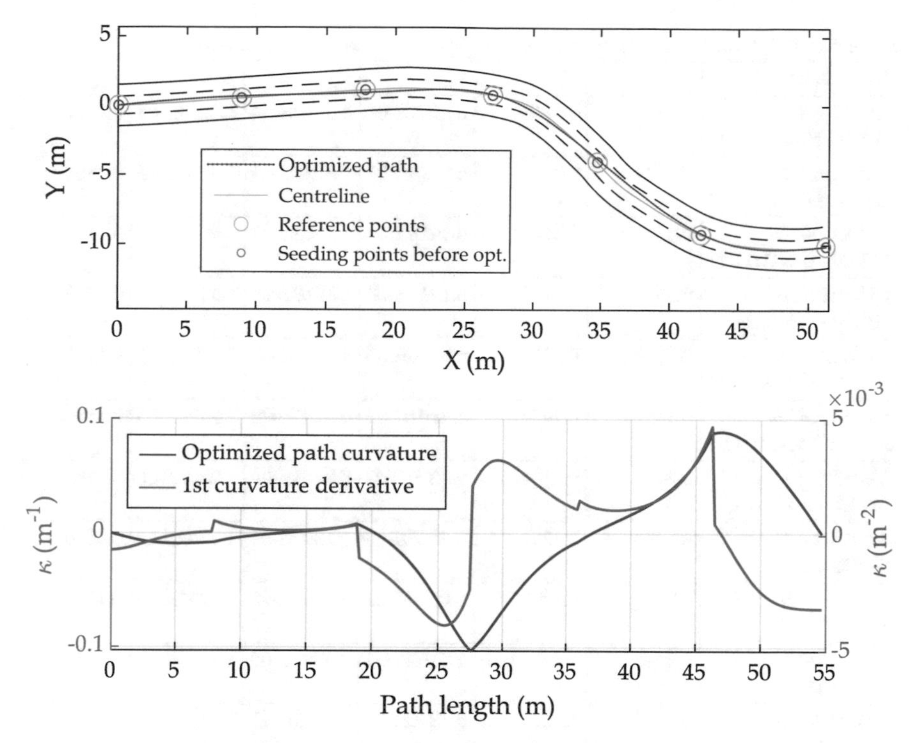

Fig. 6.14 Results of test 2:E:3:0:LL-CE-J4:00:00. Scenario and final optimized path (top) and curvature of the final path and its first derivative (bottom)

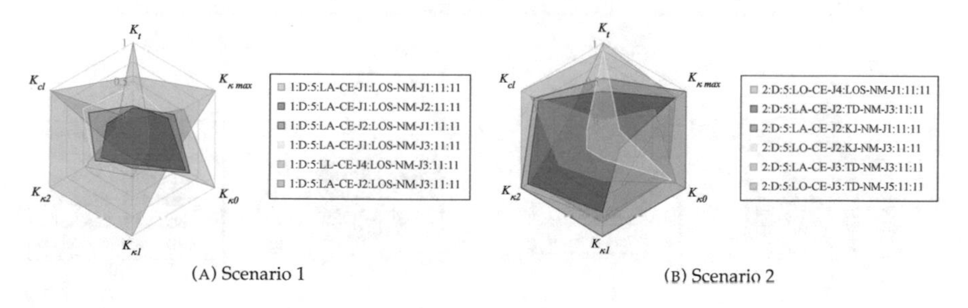

(A) Scenario 1

(B) Scenario 2

Fig. 6.15 Results of selected test cases of each scenario

The following list discusses the results obtained with respect to the different primitive configurations and optimisation techniques considered:

- **Optimization stages**: As can be noticed taking into account the results in both scenarios, the KPIs reflect a better overall performance when two optimization stages are carried out. It is also noteworthy that even the KPI K_t is lower in cases where two optimization stages were performed. Moreover, it is also remarkable that, again, the best cases with one optimization stage use quintic Bézier splines

Table 6.3 KPI values of a selection of the best tests cases from both scenarios

Test case ID	K_t	$K_{\kappa max}$	$K_{\kappa 0}$	$K_{\kappa 1}$	$K_{\kappa 2}$	K_{cl}
1:D:5:LA-CE-J1:LOS-NM-J1:11:11	0.3478	0.2146	6.3975E-02	4.3119E-04	1.8174E-05	0.1774
1:D:5:LA-CE-J1:LOS-NM-J2:11:11	0.3489	0.2127	6.1818E-02	4.3765E-04	1.8818E-05	0.1560
1:D:5:LA-CE-J2:LOS-NM-J1:11:11	0.3491	0.2115	6.1428E-02	4.3518E-04	1.9052E-05	0.1629
1:D:5:LA-CE-J1:LOS-NM-J3:11:11	0.3510	0.2116	6.1511E-02	4.3510E-04	1.9103E-05	0.1653
1:D:5:LL-CE-J4:LOS-NM-J3:11:11	0.3519	0.2207	6.0106E-02	4.8143E-04	2.0956E-05	0.1769
1:D:5:LA-CE-J2:LOS-NM-J3:11:11	0.3551	0.2077	5.8191E-02	4.4287E-04	1.9494E-05	0.1524
2:D:5:LO-CE-J4:LOS-NM-J1:11:11	0.3442	0.1164	9.5620E-03	2.8543E-05	1.2807E-06	0.1393
2:D:5:LA-CE-J2:TD-NM-J3:11:11	0.3464	0.1251	8.9540E-03	3.1399E-05	1.7702E-06	0.1380
2:D:5:LA-CE-J2:KJ-NM-J1:11:11	0.3474	0.1302	9.6590E-03	3.3555E-05	1.8731E-06	0.1388
2:D:5:LO-CE-J2:KJ-NM-J3:11:11	0.3495	0.1007	9.5130E-03	2.6601E-05	8.4680E-07	0.1297
2:D:5:LA-CE-J3:TD-NM-J3:11:11	0.3495	0.1269	9.1370E-03	3.2077E-05	1.8669E-06	0.1409
2:D:5:LO-CE-J3:TD-NM-J5:11:11	0.3501	0.1255	9.4750E-03	3.2300E-05	1.7255E-06	0.1328

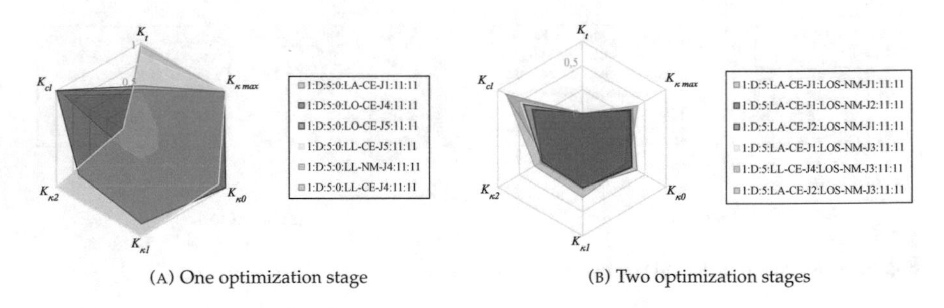

(A) One optimization stage

(B) Two optimization stages

Fig. 6.16 Results of selected test cases of Scenario 1 with one (**A**) and two (**B**) optimization stages

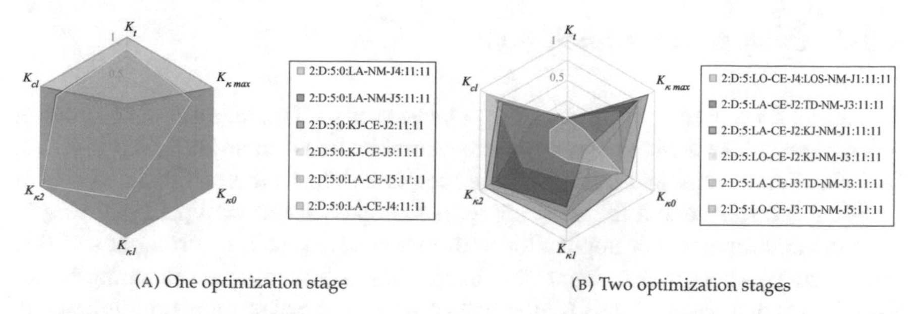

(A) One optimization stage (B) Two optimization stages

Fig. 6.17 Results of selected test cases of Scenario 2 with one (**A**) and two (**B**) optimization stages

optimized through different approaches, instead of cubic B-splines. This can be caused by the lower tunability and stability of cubic B-splines compared to quintic Bézier splines. Note that the all shown cases in these figures are the best selected after the filtering explained above. Therefore, despite the normalized value of some KPIs is close to 1, all test cases present acceptable absolute values.

- **Primitive**: Considering the results obtained, it can be concluded that some of the better approaches from those proposed in this section are those using quintic Bézier splines as primitive and two optimization stages. Furthermore, some of the better approaches for the first optimization stage are **LA, LO,** and **LL**; while for the second one are **LOS, KJ,** and **TD**.
- **Optimization methods**: Regarding optimization algorithms, SMOCE and NOMAD seem to deal better with the strong non-linearity of the optimization problem (since most of the cost function are based on the curvature equation) obtaining better overall results compared with interior-point and Levenberg-Marquardt algorithms.
- **Cost functions**: With regard to the cost functions, **J1, J2, J3,** and **J4** are present in the first optimization stage of the best results, while all of them are present in the second stage. In fact, it can be observed that some of the best results use the same configuration for the first optimization stage and different configurations in the second. This observation highlights the higher impact of the cost function in the first optimization stage when compared with the second one.

The results of all tested cases are publicly available at https://autopia.car.upm-csic.es/antonio/comparison_results.html. In this url, parallel coordinates plots as those shown in Figs. 6.7, 6.8, 6.9, 6.10, 6.11 and 6.12 can be seen. The interface allows to select the colour of the lines based on KPIs values as well as selecting the scenario and the percentile to filter data. Moreover, additional filters can be applied over the plot coordinates. In addition, the resulting KPIs values of selected test cases are shown in a table at the bottom of the web page.

6.3.4 Comparative Conclusions

An insight on a number of different approaches for path planning is carried out in this section, where a wide range of possible combinations among several primitives, optimization methods and algorithms are compared. The results are intended to help in future decisions about the most appropriate approach for local path planning in different environments or applications. To that end, the main contributions of this section are (i) a comparison framework to benchmark different path-planning primitives for on-road urban driving, (ii) the evaluation of different primitive configurations and optimisation techniques for path-planning, and (iii) the open publication of the results and its consequent analysis, based on a set KPIs related to the aforementioned main features.

6.4 Trajectory Generation

The trajectory generation is the last step of the motion planning architecture. The main goal of this module is to provide a new trajectory when requested by the manoeuvre planner, with the aim of achieving the best trade-off between optimality and planning time [2]. Note that in this document, it is referred to as trajectory the composition of a path with an associated speed profile. Bearing this in mind, the computed trajectory must meet a set of requirements:

- To ensure comfort inside the vehicle, steering and pedals behaviour must be smooth and continuous. In order words, lateral and longitudinal accelerations should not exceed specified maximum values along the trajectory.
- The trajectory generator must be able to provide feasible trajectories to avoid static or dynamic objects.
- The trajectory must be computed in a reasonable amount of time in order to be reactive enough to avoid collisions in dangerous situations.

6.4.1 Choosing the Planning Approach

Based on the above requirements, the primitive used for path planning must be able to generate a continuous curvature path. In geometric terms, that means that G^2 continuity must be guaranteed. Furthermore, the path must be computed as fast as possible, since an optimization algorithm should evaluate a large number of paths in the shortest possible time.

Taking into account the extensive comparison presented in previous subsections a quintic Bézier curve is chosen to generate the final path. Some of the main advantages of fifth order Bézier curves over cubic ones are the higher control of the curve shape and the possibility to impose curvature at both extremes of the curve. Let us recall that it is possible to concatenate quintic Bézier sections to achieve curvature continuity along the path, thus complying with comfort requirements. Moreover, these curves allows to define a wide range of curves from given initial and final poses, what allows to easily generate a number of possible paths.

Although the two-stage optimization approaches give good results in terms of curve smoothness optimization as discussed in Sect. 6.3.3.2, the computation times are generally high and indeterminate. Furthermore, approaches using quintic Bézier curves and not applying optimization algorithms have been shown to be able to find solutions with low values of the cost functions used, by evaluating a limited number of possible candidates, therefore, obtaining low and limited computation time.

In view of the above, the approach chosen for the final path generation is similar to the one with ID: **X:D:5:0:0:11:11** as defined in Sect. 6.3.2.2, where X means any scenario. However, some modifications have been made over this approach: On the one hand, the algorithm used for the centreline simplification is a modification of Douglas-Peucker algorithm that, in addition to the algorithm tolerance (ϵ_{simp}), imposes a maximum distance between two consecutive simplified points (d_{simp}^{max}) in order to extend the search space increasing the amount of **reference points** to explore. On the other hand, the **reference points** are used to create the candidates to explore by the planning algorithm instead of using them directly as waypoints. A detailed description of how this is carried out is provided in Sect. 6.4.4.

As stated, quintic Bézier curves provides a higher controllability of the curve shape over cubic ones. The higher polynomial order allows to impose the position, orientation and curvature at the extreme curve points but also two additional degrees of freedom are still available, which are used to generate a set of different curves with the same initial (p_0) and final (p_f) poses. That is achieved by varying velocity and acceleration vectors while maintaining the initial and final poses.

In order to generate a set of curves with the same orientation at their extremes, the length of the initial and final velocity vectors ($\vec{t_0}$ and $\vec{t_f}$) is varied. Firstly, to make independent the modules of the tangent vectors from each different curve cases (where the distance between extreme points is not constant), both lengths are normalized with respect to the distance between both curve extremes (d_{AB}). Then, a set of n_t points is generated between the interval $[m_t^{min}, m_t^{max}]$, where m_t^{min} and m_t^{max} are the minimum and maximum normalized lengths of the tangent vectors, respectively. Finally the length of the tangent vector is calculated as follows:

$$|\vec{T}_n^A| = |\vec{T}_n^B| = m_{tn} \cdot d_{AB} \quad \forall m_{tn} \in [m_t^{min}, m_t^{max}] \quad n = 1, \ldots, n_t \tag{6.28}$$

Furthermore, based on (6.10), the relationship between curvature and normal component of the acceleration vector is used to maintain the curvature at curve extremes while tangential component is varied to generate different curves. In that sense, the following expression is used to generate n_κ different acceleration vectors.

$$\vec{\kappa}_m = a_m \cdot \vec{T}_u^m + \kappa_m |\vec{T}^m|^2 \cdot \vec{N}_u^m \tag{6.29}$$

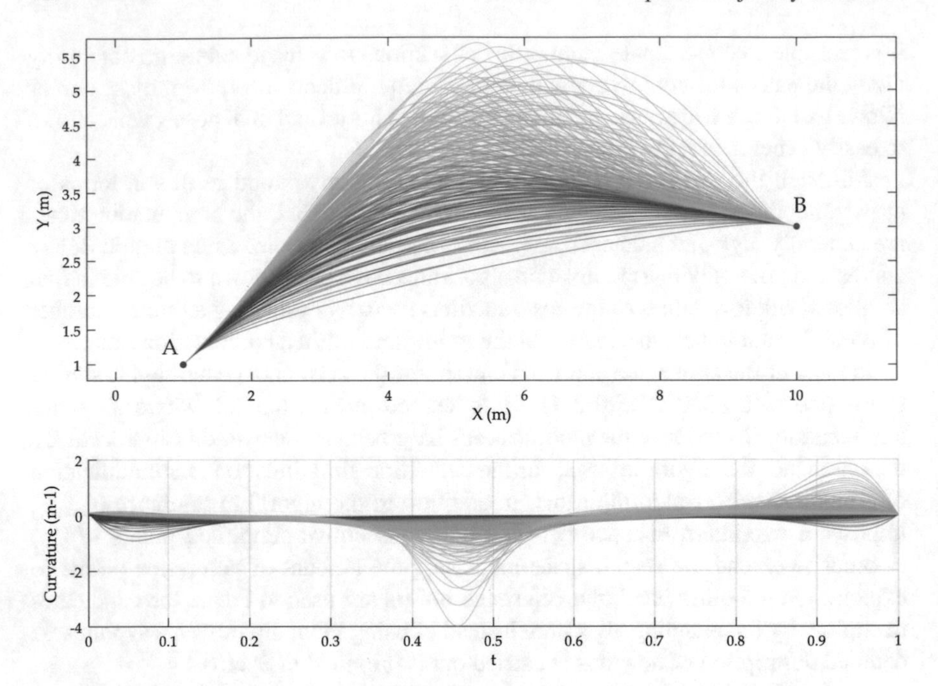

Fig. 6.18 Top: Quintic Bézier curves generated with the same initial and final poses: $A = [1, 1, 45, 0.1]$, $B = [10, 3, -10, 0]$. Bottom: Curvatures of the top curves

where $\vec{T_u^m}$ and $\vec{N_u^m}$ are the unit tangent and normal vectors at point m of the curve and, by definition, $a_t = |\vec{T^m}|'$. However, as curvature is only dependent of the normal component, n_κ values of a_m are imposed to generate different acceleration vectors with the same curvature at point m.

Just like in the case of the velocity vector length, the tangential component of the acceleration vector is modified proportionally to distance d_{AB} as expressed in Eq. (6.30).

$$a_n^m = m_{\kappa n} \cdot d_{AB} \quad \forall m_{\kappa n} \in [m_\kappa^{min}, m_\kappa^{max}] \quad n = 1, \ldots, n_\kappa \tag{6.30}$$

Once velocity and acceleration vector are calculated for two given poses, Eqs. (6.15) to (6.19) are applied to generate quintic Bézier curves imposing all combinations of velocity and acceleration vectors at both extremes. This method allows to generate a set of different curves maintaining the initial and final poses, allowing a better space exploration from the same inputs (p_0 and p_f). Figure 6.18 shows a set of curves generated with the same initial and final poses and the curvature of each curve, where a range of colours has been used to relate each curve (at the top of the figure) with its curvature (at the bottom).

6.4.2 Collision Checking

In sampling-based motion planning approaches, collision checking should be carried out for each sampled system state. Thus, collision checking is the most computationally expensive process in most of the search-based motion planning algorithms [18].

Some approaches such as those presented in [17] introduce two collision checking stages in order to firstly make a fast approximation of the possible colliding states. Then a second and more accurate collision computation is performed.

Several collision checking approaches start from the rectangular vehicle shape and then approximate this rectangle through a set of circles (typically 3, 4, 6 or 8 circles) [17, 18]. The main motivation for these approaches is the low computing time of the collision checking as only the computation of euclidean distances is needed. The main drawback of these approaches is the loss of accuracy when few circles are used to approximate the vehicle shape. The key challenge is to balance the computing time and collision checking accuracy.

Instead of circle-based approximations of the vehicle footprint, the approach for collision checking presented in this thesis uses the bounding rectangle of the vehicle. The method is based on the generation of a polygon that represents the space that the vehicle would take while driving along the calculated path. Finally, this occupancy polygon is used to firstly check if the path is inside the road corridor and then if any obstacle collides with it.

In order to obtain the occupancy polygon, the dimensions of the vehicle and the path generated as explained in Sect. 6.4.1 are needed. Taking advantage of the fact that the path is a Bézier curve or a concatenation of them, the tangent vector and the curvature can be obtained analytically.

Based on the path, the right and left bounds of the area occupied by the vehicle can be calculated as follows: the right bound will be composed of the points of right extreme of the front of the vehicle when the vehicle is turning left and of the points of right extreme of the rear axle when turning right. The left bound is calculated analogously: it is composed of the points of left extreme of the front of the vehicle when the vehicle is turning right and of the points of left extreme of the rear axle when turning left.

It is worth to mention that a safety margin is added around the vehicle (d_{sm}). To determine if vehicle is turning right or left, the sign of the curvature is used. Finally, the polygon is conformed by joining the points of right and left sides to obtain a closed shape. Figure 6.19 shows an example of occupancy polygon for a given path, where l_{tw} is the vehicle axle track, l_{la} is the distance from the rear axle to the front bumper and l_{lb} is the distance from rear bumper to the rear axle.

In point B of Fig. 6.19 can be seen how the vehicle is turning right and the extreme left point of the front belongs to the left bound of the polygon, while the right extreme point of the rear axle y used for the right bound.

Once the path occupancy polygon is obtained, it is used to check if it is inside the road corridor and also if it collides with some obstacle. The verification strategies for both cases rely on the algorithm 5 described in [12] to solve the "point in polygon"

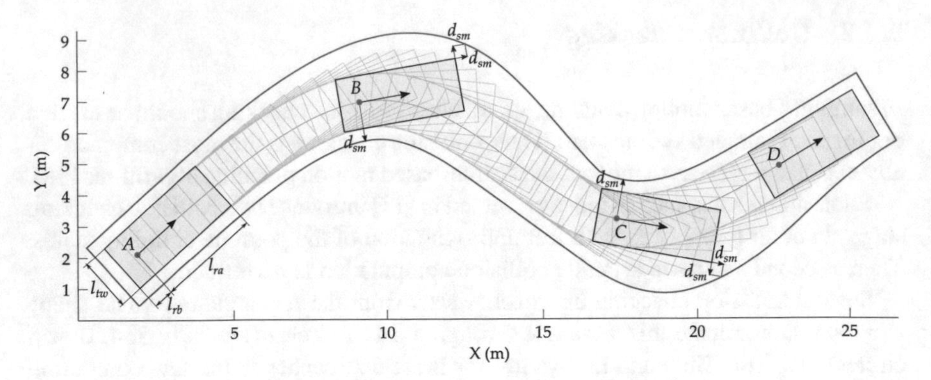

Fig. 6.19 Example of the path occupancy polygon calculation for collision checking

problem. However, our approach introduces several modifications to the original version.

Verify if the path polygon is inside a given road corridor It is checked that all vertexes of a simplified path polygon are inside the road corridor. When a vertex of the path polygon is outside the road corridor polygon, the execution stops.

Verify if the path polygon collides with some obstacle In this case all vertexes of the obstacles are checked to be outside the path polygon. When a vertex of any obstacle is inside the path polygon the execution stops as in the previous case.

6.4.3 Speed Profile Generation

The speed profile is calculated over a given path so that a longitudinal speed value is associated to each of its points.

In order to comply with requirements, the generated speed profile must limit both longitudinal and lateral accelerations as well as maximum speed to comply with traffic rules and to ensure comfort inside the vehicle (see Table 6.4). In this connection, initial and final speed must be also imposed.

Since the road corridor and the centreline are composed of Bézier curves as described in Sect. 4.4, the curvature in each of the simplified points over the centreline is analytically calculated when a new corridor is generated. The curvature at each reference point (κ_{R_n}) is then used to compute a maximum speed value ($v_{R_n}^{max}$) by limiting the lateral acceleration ($a_{max,lat}$) as follows:

$$v_{R_n}^{max} = \sqrt{\frac{a_{max,lat}}{|\kappa_{R_n}|}} \tag{6.31}$$

These maximum speed values at reference points will be used to set the final speed of the final trajectory computation:

Table 6.4 Speed profile generation parameters

Symbol	Description
v_0	Initial speed
v_f	Final speed
v_{max}	Maximum allowed speed
$a_{max,lat}$	Maximum lateral acceleration
$a_{max,acc}$	Maximum positive longitudinal acceleration
$a_{max,dec}$	Maximum negative longitudinal acceleration

$$v_f = v_{R_n}^{max} \tag{6.32}$$

where R_n is the reference point over the centreline used to generate the candidate selected to generate the final trajectory).

The speed profile calculation is carried out in several stages:

1. Firstly, a speed limit curve is computed based on maximum lateral acceleration allowed ($a_{max,lat}$). To that end, the speed limit at each point of the path is computed considering circular motion with the curvature:

$$v_{n,limit} = \sqrt{\frac{a_{max,lat}}{|\kappa_n|}} \tag{6.33}$$

   where n is the index of path point. Then the maximum speed at point n is calculated as follows:

$$v_n = \min\{v_{max}, v_{n,limit}\} \tag{6.34}$$

2. After that, longitudinal accelerations are limited. To do that, initial and final speeds (v_0 and v_f) are imposed and the acceleration profile is computed assuming uniform acceleration between two consecutive points of the path, as shown in Eq. (6.35).

$$v_n = \sqrt{v_{n-1}^2 + 2a_n d_p} \tag{6.35}$$

   where d_p is the distance between points $n-1$ and n of the path.
3. Then, the accelerations computed for each path point are traversed forward in order to verify that it is lower than the maximum acceleration value ($a_{max,acc}$). In case that the acceleration at point n overcomes the limit, it is thresholded to the maximum value and the speed at point $n+1$ is recalculated using Eq. (6.35).
4. Finally, the same procedure followed in the previous step is performed backwards imposing a deceleration limit of a_{dec} along the whole path.

As an example, Fig. 6.20 shows two speed profiles computed for the same path with different values of the parameters of Table 6.5. The maximum lateral acceleration shown in this table ($a_{max,lat}(m/s^2)$) has been obtained by considering 30% of

Table 6.5 Speed profile parameters used in examples of Fig. 6.20

Parameter	Speed profile 1	Speed profile 2
v_{max} (km/h)	30	30
$a_{max,lat}$ (m/s^2)	1.04	1.04
$a_{max,acc}$ (m/s^2)	1.12	0.56
$a_{max,dec}$ (m/s^2)	2.40	1.20

the lateral acceleration applied to the vehicle when driving at 15 km/h through a tight curve with a turning radius of 10 metres. The maximum longitudinal accelerations ($a_{max,acc}$ (m/s^2)) have been calculated, as the 40% (speed profile 1) and 20% (speed profile 2) of a the acceleration applied to the vehicle to accelerate from 0 km/h to 100 km/h in 10 s. Finally, the imposed maximum decelerations are the 60% and 30% of a maximum braking acceleration of 4 m/s^2, respectively.

As can be seen, the **speed profile 1** is calculated by using higher values of longitudinal accelerations. Consequently, greater speeds are achieved along the same path, as shown at the bottom of Fig. 6.20.

When a replanning is being performed, the initial speed is set to the current vehicle speed and the final speed is taken from the speed estimation of the road corridor centreline performed, as described in Sect. 6.4.3, when a new road corridor is set. It is important to emphasize the importance of speed imposed at the end of the trajectory, as it ensures the anticipation of the speed calculation over a path taking into account the road corridor features beyond where the current planed path ends.

It is also remarkable that this approach allows to change the speed planner parameters ($a_{max,lat}$, $a_{max,acc}$, $a_{max,dec}$ and v_{max}). That means that it is possible to adapt the speed behaviour over the path based on passengers needs allowing them to choose the driving abruptness level.

As depicted in Fig. 6.21, a new speed profile is needed when a moving obstacle moves perpendicularly to the trajectory of the vehicle. In order to choose a speed profile that avoids the temporal collision with the obstacle a set of speed profiles are calculated by decreasing iteratively the maximum speed in the section from the current position vehicle to the collision point of the path, maintaining the stated maximum speed (v_{max}) for the speed profile of the remaining path. Thus, when a speed profile complies with $\Delta T_{coll} > T_{th}$ in the spatial collision point, the speed profile of the current trajectory is updated to avoid the collision, and then, after the vehicle reach the collision point, the speed profile calculation is resumed.

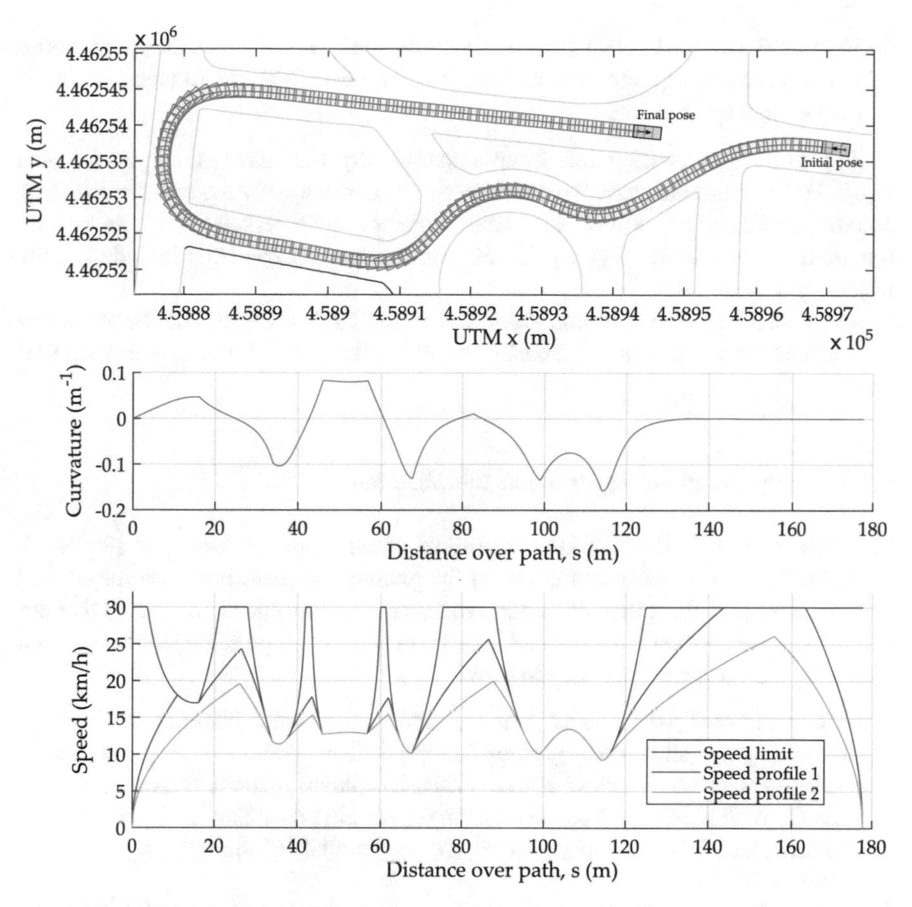

Fig. 6.20 Top: Example path. Middle: Curvature of the path. Bottom: Two speed profiles computed over the same path

6.4.4 Trajectory Generation Algorithm

The main goal of the trajectory generation module is to provide a new trajectory when requested by the manoeuvre planner. The proposed strategy to solve the motion planning problem comprises several steps:

1. **Motion planning problem initialization**: At this first stage, the motion planning solver defines the search space to explore depending on the planning mode that has been set.
2. **Candidates evaluation**: At this stage all the path candidates are evaluated by checking their validity and calculating their costs based on previously defined cost functions.
3. **Candidate selection**: Among the valid evaluated candidates, one is selected based on their costs values.

4. **Final trajectory calculation**: Once the best candidate is chosen, the speed profile is calculated taking into account the maximum accelerations to ensure comfort inside the vehicle.

This algorithm is triggered when requested by the manoeuvre planner (see Fig. 3.4). Since the manoeuvre planner module is continuously analysing the current situation (collision checking of the current trajectory with perceived objects, and the remaining trajectory length), a prioritization mechanism has been included in the trajectory generation.

The general motion planning algorithm is depicted in Fig. 6.21. The following subsections provides detailed descriptions of all the algorithm stages enumerated above.

6.4.4.1 Motion Planning Problem Initialization

The initialization is the first task performed when a new trajectory is requested. Its main function is to set up the rest of the planning process based on the current vehicle state and the planning mode requested, as introduced in Sect. 5.5. Four different planning modes are defined in order to address all possible situations when analysing the obstacles that are perceived:

Plan from current pose: This planning mode is used when planning for first time or in emergency situations e.g. when the vehicle is not being able to follow current trajectory with small control errors. When this planning mode is used, the initial pose for path planning is set to the current vehicle pose. The final poses for the candidates of this trajectory section are determined by the next n_{rp} reference points.

Extend current trajectory: If the trajectory generator is called with this planning mode, a point placed at the 90% of the current trajectory length is taken as the initial point for the new trajectory section. As in the previous case, the final poses for the candidates are determined by the next n_{rp} reference points.

Avoid static obstacle: In order to state the **reference points** for the candidates when a collision of the current trajectory with static obstacle is detected, the free distance from the obstacle to both boundaries of the road corridor (left and right boundaries) is measured to determine if the vehicle is able to pass the obstacle by none, one of both sides. If there is no space enough to avoid the obstacle it is assumed that the lane is locked and the trajectory is shortened to stop ahead the obstacle maintaining a safe distance, and a new speed profile is calculated with $v_f = 0$ km/h. In this case, an alternative route to reach the same destination is requested to the global planner. In the cases that the vehicle is able to avoid the obstacle, a set of equidistant **reference points** is computed over the free space of the perpendicular line to the centreline of the road corridor that pass through the centre of the obstacle as shown in Fig. 6.22.

The point of the current path placed at a distance d_{static} (m) from the obstacle is set as initial point of the new trajectory section. In the case that the distance from

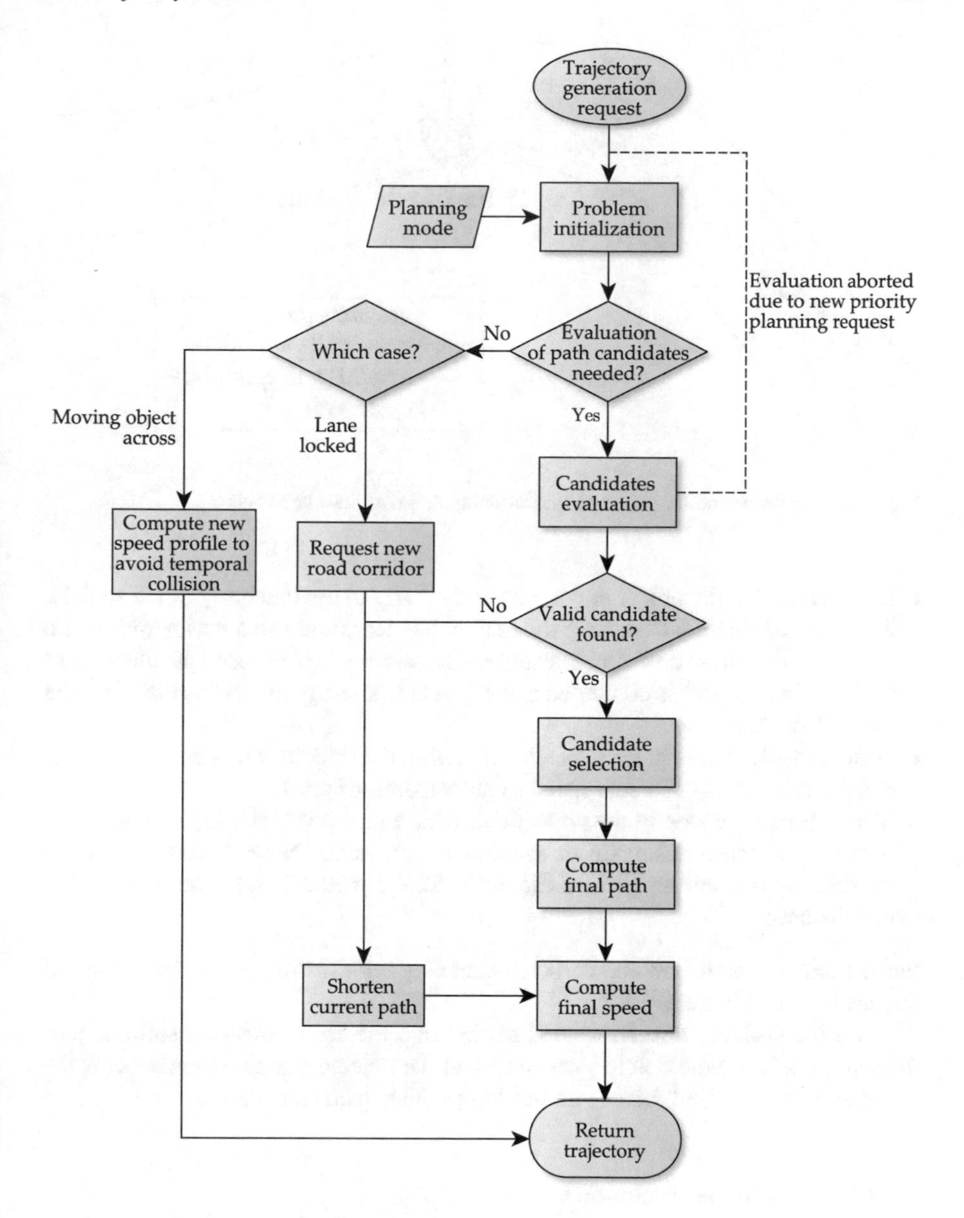

Fig. 6.21 General flow diagram of motion planning algorithm

the vehicle to the obstacle is lower than d_{static}, the closest point of the current path to the vehicle is taken as the initial point for candidates.

Avoid dynamic obstacle: If a future collision with dynamic obstacle is detected, the motion direction of the obstacle is used to determine the strategy to avoid it:

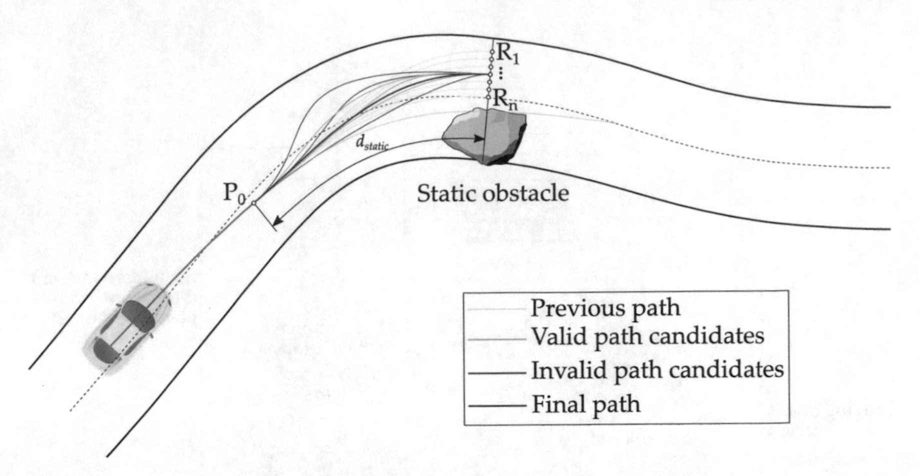

Fig. 6.22 Proposed candidates in case that an static obstacles must be avoided

- In the event that the object moves perpendicularly to the trajectory of the vehicle, it is assumed that it will outside the road in the near future so a new speed profile that avoid the future collision is searched. In case that a new speed profile can not be found, a new path is computed using the same strategy that is described for the static obstacles above.
- If the obstacle moves in the same direction that the vehicle, a new speed profile is computed using the obstacle speed as the maximum speed.
- If the obstacle moves in the opposite direction to the vehicle, the strategy used is similar to those described to avoid static obstacles. Nevertheless, instead of considering the current pose of the obstacle, the predicted pose in the collision point is used.

Further details about how the collision checking with moving obstacles is carried out can be found in Sect. 6.4.2.

Once the problem initialization is carried out, the candidates evaluation is performed if needed. Note that in cases such that the trajectory is shortened or only the speed profile is updated there is no need to evaluate path candidates.

6.4.4.2 Candidates Evaluation

In this stage, all the path candidates that have been selected in the problem initialization are computed and evaluated.

The first step in the evaluation of a candidate is to check its validity i.e. check if the path can be driven by the vehicle i.e. if the maximum curvature of the path candidate is below the maximum feasible curvature by the car ($\kappa_{max}^{p_c} < \kappa_{max}^{v}$, see Fig. 6.23A), if it would lead the vehicle outside the road corridor (see Fig. 6.23B) or if it would lead the vehicle to collide with some obstacle (see Fig. 6.23C). If the

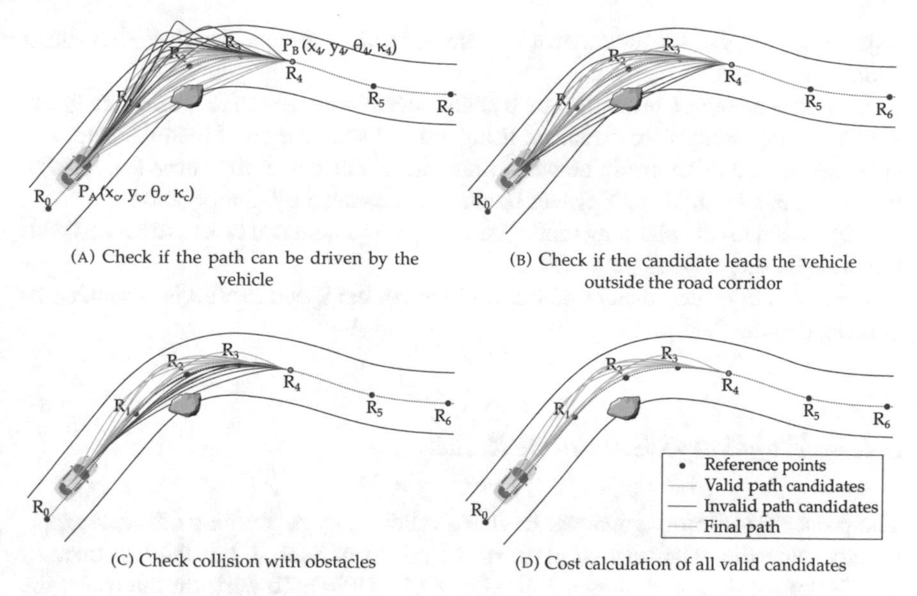

(A) Check if the path can be driven by the vehicle

(B) Check if the candidate leads the vehicle outside the road corridor

(C) Check collision with obstacles

(D) Cost calculation of all valid candidates

Fig. 6.23 Candidates evaluation steps

After the validity of the path candidate is satisfactorily checked, if it is valid, the cost function used to evaluated the quality of the path is computed (see Fig. 6.23D). As described in Sect. 6.3, it is not a trivial task to choose a cost function that define the goodness of a path in a given context. Nevertheless, based on the comparison carried out, the following cost function is used in the evaluation of the candidates.

$$J_p = \frac{1}{w_{L_p} L_p} \int_{s_0}^{s_f} \dot{\kappa}(s)^2 + w_{\ddot{\kappa}} \ddot{\kappa}(s)^2 \, ds \tag{6.36}$$

This cost function is similar to J_3, which was used in the primitive comparison. On the one hand, first and second derivatives of the curvature reflect the smoothness of the path along the curve. Moreover, the length of the path (L_p) is used to normalize its result. The main difference with J_3 relies on the use of the weight parameter w_{L_p}. The motivation for adding this weight is that in case that $w_{L_p} = 1$ (i.e. this weight is nor considered), the path that minimizes the cost functions tends to be straight and short in spite of being in curved road sections, thus obtaining almost straight paths in cases in which it should not. In order to avoid that, w_{L_p} is used with values greater than 1.

6.4.4.3 Best Candidate Selection and Final Trajectory Calculation

Among all valid candidates evaluated in the previous stage, the candidate with minimum cost is selected as the path for the final trajectory. In order to obtain the final trajectory, firstly the Bézier curve of the selected path candidate is evaluated to obtain equidistant points.

Equidistant points can not be directly computed in a Bézier curve. To do that, firstly the curve polynomials are evaluated using a fine discretization of the parameter t to approximate the relationship between t and the distance over the curve (s). Finally, the values of t to obtain equidistant points are calculated by interpolation.

Depending on the planning mode, the new path section can be concatenated with a section of the previous path or not.

Once a path of equidistant points is obtained, the speed profile is computed as described in the Sect. 6.4.3.

6.4.5 Trajectory Generation Results

The proposed trajectory generator has been validated by performing different experiments using the experimental platform described in Sect. 7.2 at the test track of the Centre for Automation and Robotics (CSIC-UPM). To perform the trials, the trajectory generator has been integrated in the architecture proposed in Chap. 3. Nevertheless, this subsection focuses in the results of the trajectory generator.

The algorithm is evaluated in two different driving scenes: (i) an urban-like scenario with sharp curves and (ii) an urban-like scenario where static and dynamic obstacles must be avoided.

In the experiments carried out, the trajectory generator was set up to compute a total of 4500 path candidates in each planning request. The detailed list of parameters and values used in these experiments are shown in Tables 6.6 and 6.7.

Table 6.6 Path planning configuration

Parameter	Description	Value
d_{sm}	Safety distance around vehicle (m)	0.4
min_{pl}	Minimum trajectory length threshold (m)	55
ϵ_{simp}	Tolerance for centreline simplification (m)	0.25
d_{simp}^{max}	Maximum distance between reference points (m)	7
n_{rp}	Number of reference points used to create candidates	15
n_t	Initial and final tangent vector magnitude evaluation number	10
n_κ	Initial curvature vector magnitude evaluation number	3
m_t^{min}	Minimum normalized tangent vector magnitude	0.3
m_t^{max}	Maximum normalized tangent vector magnitude	1.7
m_κ^{min}	Minimum normalized curvature vector magnitude	0
m_κ^{max}	Maximum normalized curvature vector magnitude	10

Table 6.7 Speed planning configuration

Parameter	Description	Value
v_{max}	Maximum allowed speed (km/h)	20
$a_{max,lat}$	Maximum lateral acceleration (m/s^2)	1.0
$a_{max,acc}$	Maximum positive longitudinal acceleration (m/s^2)	0.4
$a_{max,dec}$	Maximum negative longitudinal acceleration (m/s^2)	0.7

6.4.5.1 Scenario 1: Urban-Like Route Through Tight Curves

This scenario presents a highly sharp road corridor in which the trajectory generator is tested. The route includes the entrance and exit of a small roundabout and several 90° curves in a single narrow lane of 5 meters wide approximately. Figure 6.24 shows the valid candidates evaluated in different planning request during the trial. Notice that invalid candidates has not been plotted in this sub-figures and still the trajectory generation algorithm can choose the final candidate among a number of valid candidates that evaluated in each planning request.

Figure 6.25 shows the path seen by the vehicle controllers and the real vehicle path during the performed trial. As can be visually noticed, the whole path presents a smooth shape similar to typical vehicle paths obtained when a human is driving.

Figure 6.26 shows information of the trajectory tracking during the full trial in scenario 1. In addition to the visual perception of the path smoothness of Fig. 6.24 and 6.25, in Fig. 6.26A it can be seen that both lateral and angular errors used in the lateral control present a smooth behaviour, even though the trajectory is updated 13 times during the trial.

Moreover, Fig. 6.26D shows the instant and mode of the planning requests performed. It can be noticed that the mode of the first planning request is *0*, corresponding to plan from the current vehicle pose as there no initial trajectory. Since this scenario does not comprises obstacles, the rest of the planning requests are performed in mode *1* (extend current trajectory). In addition, it can be observed that the first four planning requests are consecutively performed. This is caused by the minimum remaining path length requirement, that makes the manoeuvre planner to send planning requests to the trajectory generator with planning mode *1* (extend current trajectory) until this requirement is met.

Finally, Fig. 6.26C shows the reference speed and measured vehicle speed during the test.

Regarding the computing time, the mean of the processing time per planning request is 56.37 ms with and standard deviation of 14.69 ms. Furthermore, more information of each planning request is shown in Table 6.8.

Regarding the speed planning, the maximum positive longitudinal, negative longitudinal and lateral accelerations were set to 0.4 m/s^2, 0.7 m/s^2 and 1.0 m/s^2, respectively. To analyze the resulting behaviour of the vehicle in terms of occupant comfort, Fig. 6.27 shows a density plot of the real longitudinal and lateral accelerations to mea-

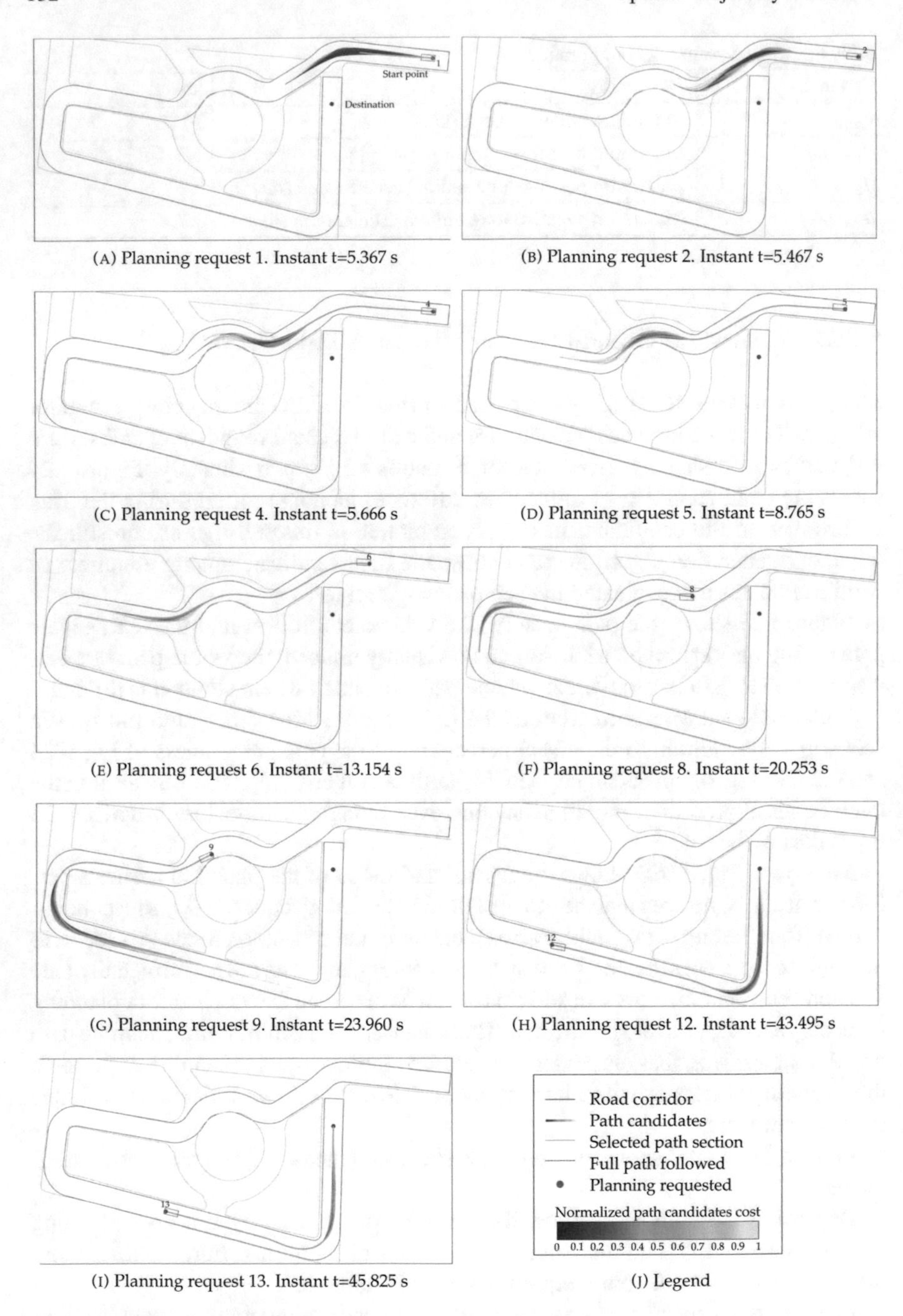

Fig. 6.24 Valid and selected candidates at some of the planning requests during the trial in scenario 1

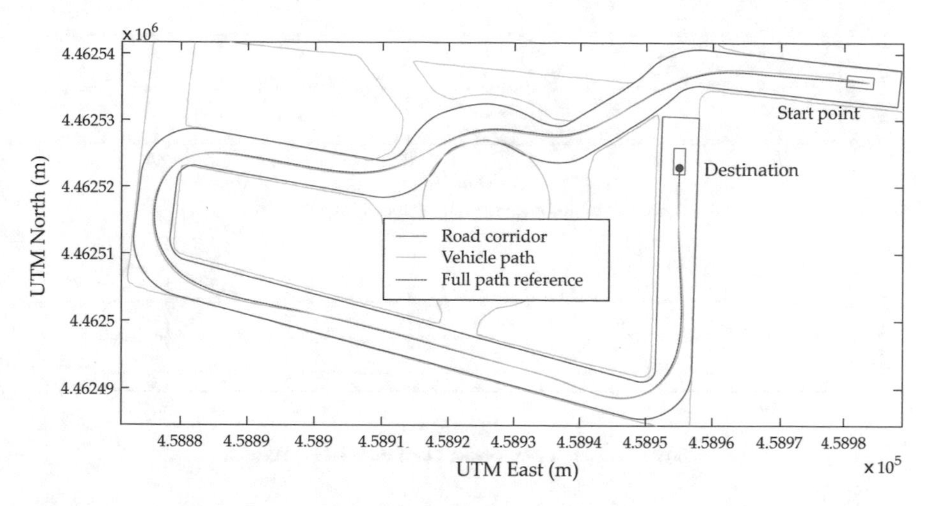

Fig. 6.25 Final reference path and vehicle path in the scenario 1 trial

Table 6.8 Additional information related to relevant planning requests during the trial in scenario 1

Planning request ID	Request time-stamp (s)	Planning mode	Planning time (ms)	Valid candidates (%)
1	5.37	0	73.96	33.51
2	5.47	1	64.43	21.80
3	5.56	1	61.04	16.11
4	5.67	1	65.50	12.38
5	8.77	1	64.24	5.93
6	13.15	1	52.98	3.44
7	16.61	1	57.44	17.27
8	20.25	1	51.61	4.04
9	23.96	1	58.29	7.04
10	32.33	1	76.77	35.11
11	42.15	1	42.90	9.80
12	43.50	1	42.01	12.00
13	45.83	1	21.61	14.42

sured along the trial. This figure depicts that most of the acceleration measurements fall within the dashed white rectangle that represent the stated acceleration limits.

To conclude, this experiment showed that the proposed algorithm is able to generate a number of valid candidates and select the optimum candidates in a few milliseconds even in sharp areas where consecutive curves must be overcome by the vehicle.

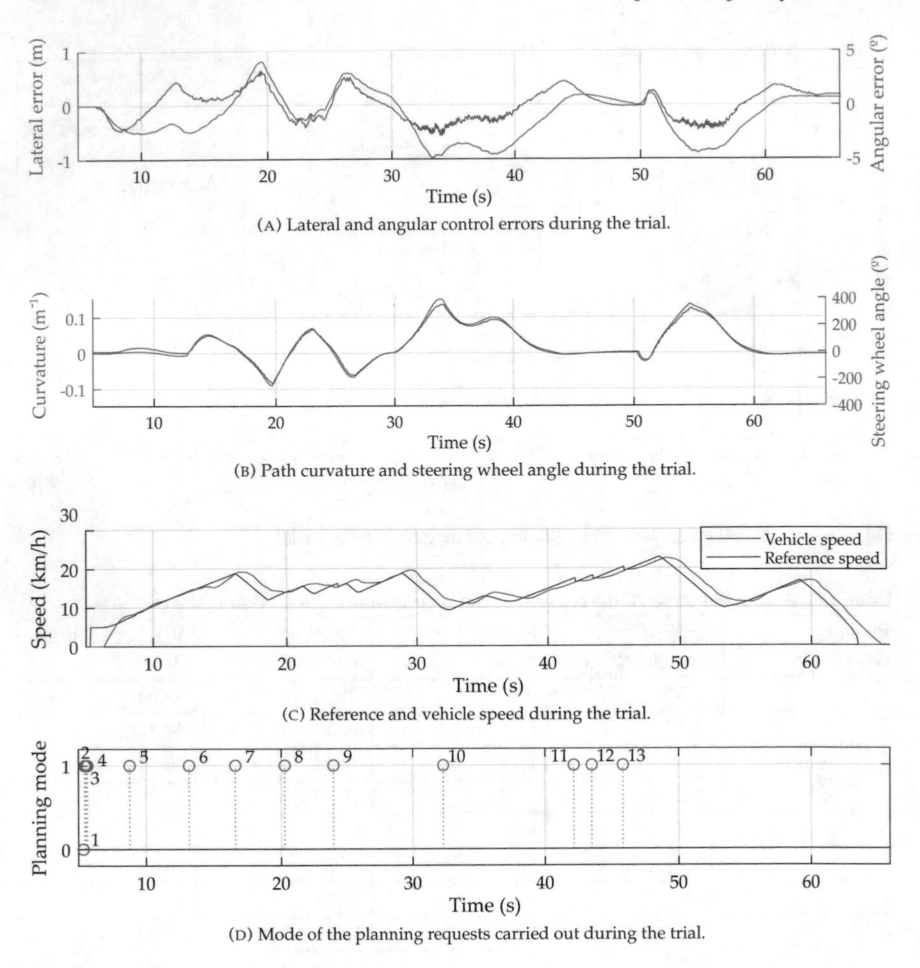

(A) Lateral and angular control errors during the trial.

(B) Path curvature and steering wheel angle during the trial.

(C) Reference and vehicle speed during the trial.

(D) Mode of the planning requests carried out during the trial.

Fig. 6.26 Trajectory tracking in scenario 1 trial

6.4.5.2 Scenario 2: Static and Dynamic Obstacles Avoidance

The scenario of this experiment includes static and dynamic obstacles located at different places of the route that the vehicle is following to reach the destination point, which was set a few meters behind the initial point as can be seen in Fig. 6.28. As this scenario includes a greater complexity with respect to the previous one, and consequently a higher number of planning request is carried out, only the valid path candidates of five representative planning requests (*1, 6, 16, 40, 42*) during the trial are plotted to improve the readability of this figure. To complement the understanding of this plot, Table 6.9 shows relevant information about the representative planning requests of Fig. 6.28. In this table the planning modes correspondence is *0* - re-plan from current pose, *1* - extend current trajectory, *2* - avoid static obstacle, and *3* - avoid dynamic obstacle. Moreover, the percentage of valid candidates and the processing

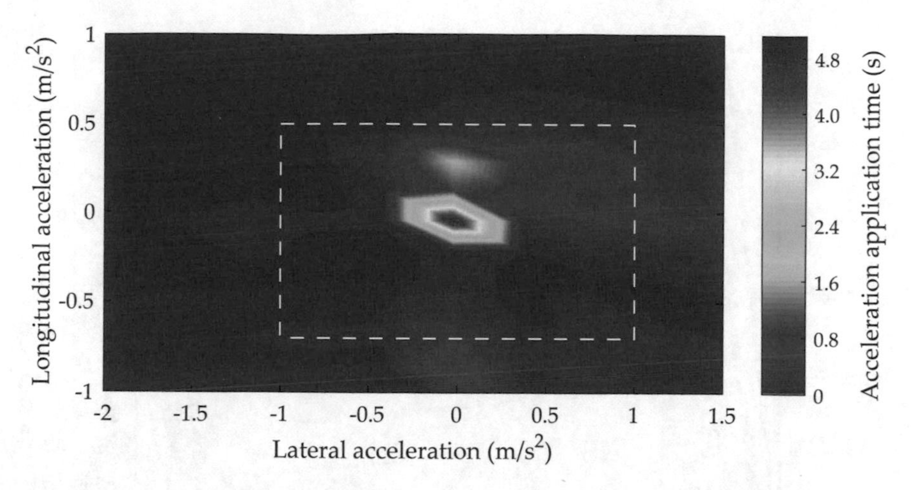

Fig. 6.27 Density graph of measured acceleration in the vehicle during the trial in scenario 1

time in milliseconds at each request are also shown in this table. The average planning time for the whole trial was 37 ms with a standard deviation of 25 ms for the whole experiment.

Furthermore, Fig. 6.29 shows the speed profile and vehicle speed (left ordinate axis) together with the planning requests during the experiment (right ordinate axis). In this figure it can be seen how and when the planning requests are triggered with different planning modes depending on the situation: at the beginning, the first trajectory is planned from the current vehicle pose (mode *0*). At instant t = 31.47 s, a dynamic obstacle is detected and consequently a new planning request is carried out (request ID 10). Later, at instant t = 45.54 s, the first static obstacle is detected and a planning request of mode 2 is sent (request ID 16). Afterwards, successive planning requests are closely triggered (request IDs 17–42). This is caused by the small shape and localization changes of the perceived obstacles at different consecutive instants, which leads to launch the candidates evaluation procedure at a higher frequency. Note that the trajectory is quickly corrected to avoid the obstacles satisfactorily even with highly noisy perception information. It is also to be noticed that although the speed profile is smooth and continuous almost at every moment, there are two instants (planning requests 10 and 11), where the sudden incursion of the pedestrian in the road forces to reduce the target speed so that the vehicle keeps in the safe envelope.

In order to analyse the comfort inside the vehicle during the test, Fig. 6.30 shows a density map to represent the measured longitudinal and lateral accelerations. In these tests, the maximum positive longitudinal, negative longitudinal and lateral accelerations were set to 0.4 m/s^2, 0.7 m/s^2 and 1.0 m/s^2, respectively.

Figure 6.30 shows how most of the measured acceleration values fall within the limits (marked with a dashed white rectangle in Fig. 6.30) established in the planning. However, some values are outside mainly due to the joint effect of vibrations induced by road imperfections and road and vehicle pitching and rolling. It can also be noticed

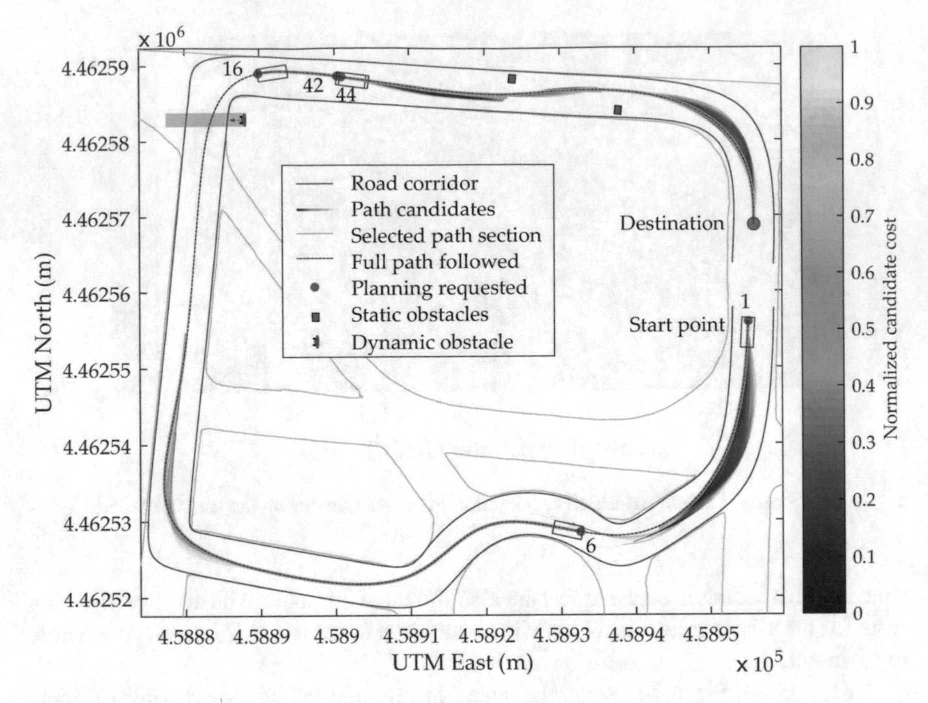

Fig. 6.28 Resulting paths in scenario 2

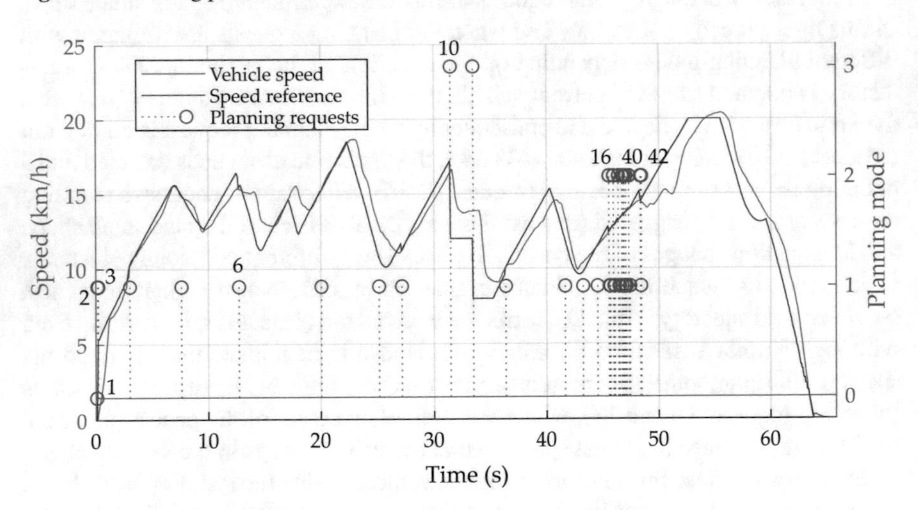

Fig. 6.29 Resulting speed profile and planning requests in scenario 2

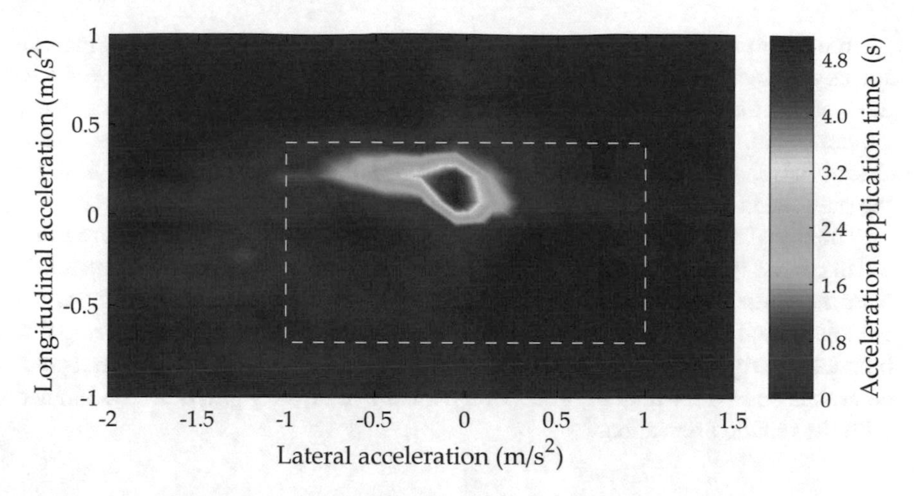

Fig. 6.30 Density graph of measured acceleration in the vehicle during the trial in scenario 2

Table 6.9 Additional information related to relevant planning requests during the trial in scenario 2

Planning request ID	Request time-stamp (s)	Planning mode	Planning time (ms)	Valid candidates (%)
1	0.092	0	54.45	30.18
6	12.70	1	36.66	33.22
10	31.47	3	1.01	–
16	45.54	2	67.23	34.24
40	48.34	2	70.40	27.56
42	48.43	2	14.49	41.47

that since the vehicle must go through more right turns than left ones to reach the destination, more negative (acceleration applied in left direction) than positive lateral acceleration values are measured.

6.5 Occupancy Grid-Based Motion Planning

As described in Sect. 4.6, when maps are used as a part of the environment understanding, the good localization with respect to the map becomes critical. To deal with localization uncertainty in motion planning when map data is used, the grid-based approach presented in Sect. 4.6 has been implemented. By means of this approach the probability of occupation of each grid cell is obtained, thus providing more information about the occupancy of the nearby environment that merely the road corridor as considerer so far.

In order to use the occupancy probability information in the local planning, the stages **problem initialization** and **candidates evaluation** (described in Sect. 6.4.4.2) have been extended. As described in Sect. 6.4.4.1, this first stage of the trajectory generator is in charge of defining the search space to create the path candidates. Taking into account the size of the occupancy grid, the **problem initialization** states the path candidates from the reference points that falls within the grid.

The size of the occupancy grid is updated every time it is computed. This is carried out to cover a specified amount of meters ahead the current vehicle position measured over the centreline of the road corridor. In this context, the road corridor ahead the vehicle is used to calculate the width and height of the occupancy grid ensuring that both left and right boundaries from the vehicle position until the last considered point centreline point fall within the grid. Moreover, the occupancy grid is always aligned with the vehicle orientation.

6.5.1 Candidates Evaluation with Occupancy Grid

The main difference when using the occupancy grid for motion planning with respect to the general approach presented in Sect. 6.4.4.2 is found in the **candidates evaluation** stage. In this case, instead of having the drivable space boundaries as input for the evaluation the validity of each path candidate and the path cost related to its smoothness, the occupancy grid is the information baseline to compute both the candidates validity and cost.

The validity is determined similarly than in the general case i.e. by checking the three verifications: firstly (i) the maximum curvature of the path candidate ($k_{max}^{p_c}$) must be lower than the maximum curvature feasible by the vehicle (k_{max}^{v}); after that, (ii) it is checked if the path is inside the road corridor and also (iii) if there is a collision with any obstacle.

One of the main advantages of the occupancy grid approach is that it allows to fuse environmental information coming from different sources. In this case, both the road corridor and the static objects perceived by exteroceptive sensors are rasterized into the grid. This method allows to abstract the motion planning algorithm to the different information sources, focusing only in one occupancy grid. That means that points (ii) and (iii) mentioned in the paragraph above can be verified at once. Therefore, instead of using the algorithm proposed in [12] to check if the path is inside the road corridor and check possible collision with static obstacles, the occupancy values of grid cell occupied by the candidate are used to accomplish this.

The path occupancy polygon of the candidate is computed just as in the general case (see Sect. 6.4.2). However, once it has been verified that the maximum curvature verification has been passed, the cells that the candidate occupies are found. To do that, firstly the candidate polygon is rasterized on a empty grid with the same size that the occupancy grid and the scan-line flood fill algorithm is used to find the cells that fall within the polygon. Thus, a list of all cells occupied by the candidate is obtained and can be used to extract occupancy data from the occupancy grid to verify its

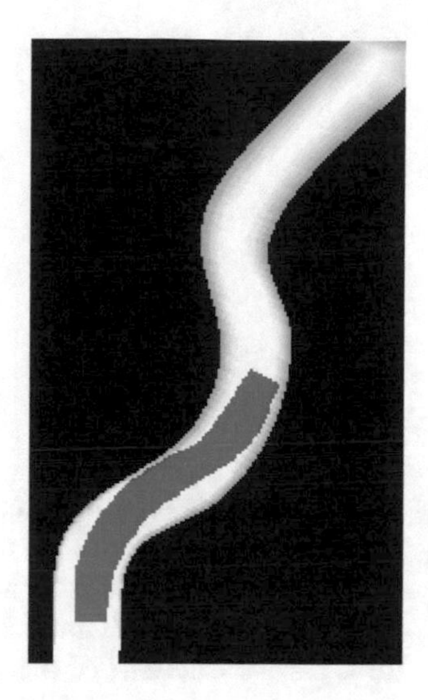

Fig. 6.31 Example of path-polygon rasterization over an occupancy grid

validity and compute its cost. An example of the rasterization of a path-polygon is shown in Fig. 6.31.

To determine the validity of the candidate, the occupancy probability of each cell occupied by the candidate must be below a given occupancy probability threshold (Pv_{th}). Thus, the candidate is valid if the following equation is verified:

$$P^c_{max} < Pv_{th} \tag{6.37}$$

where P^c_{max} is the maximum occupancy of the cells occupied by the candidate.

Once the validity of the candidate is checked, its cost is computed. The cost function used in this case is similar to that used in the general approach, according to Eq. (6.36). Nevertheless, a new component has been added to the function to represent the mean occupancy probability over the path candidate ($\overline{P}_{pc}$). Moreover, w_{og} is used to weight the occupancy mean over the path with respect the other function components:

$$J_p = w_{og}\overline{P}_{pc} + \frac{1}{w_{L_p}L_p}\int_{s_0}^{s_f} \dot{\kappa}(s)^2 + w_{\ddot{\kappa}}\ddot{\kappa}(s)^2\, ds \tag{6.38}$$

6.5.2 *Motion Planning Results Using the Occupancy Grid*

Different trials has been carried out in three different scenarios to test the proposed motion planning algorithm based on occupancy grid. The same scenarios used to test the trajectory generator in Sect. 6.4.5 are used to test the adapted motion planning algorithm using the occupancy grid (scenario 1 and 2). Finally a third scenario that with a longer route has been tested.

Table 6.10 Occupancy-grid parameters

Parameter	Value
Distance ahead (m)	60
Grid cell size (m)	0.2

Table 6.11 Considered localization uncertainties

Parameter	Value
σ_x (m)	0.02
σ_y (m)	0.02
σ_θ (rad)	0.05

The grid-based approach required the specification of new parameters: on the one hand, those used to define the occupancy grid properties are shown in Table 6.10. On the other hand, the localization uncertainties used to apply the algorithm proposed in Sect. 4.6 are shown in Table 6.11. The values of both tables have been the used for the three experiments carried out. Regarding trajectory generation algorithm, the same parameters used in the trials presented in Sect. 6.4.5 are applied in the subsequent experiments, with the exception of min_{pl}, that has been set to 50 m instead 55 m. This parametrization have been selected empirically in order to find a balance between computing time, grid accuracy and distance ahead the vehicle.

6.5.2.1 Scenario 1: Urban-Like Route Through Tight Curves Using the Occupancy Grid

This scenario consists of an urban-like layout where the vehicle has to drive through consecutive tight curves. The details of this scenario are the same that those defined in Sect. 6.4.5.1.

Figure 6.32 shows the concatenation of all paths followed by the vehicle and the real vehicle path during the performed trial. The resulting path is similar to the one obtained in the case tested in Sect. 6.4.5.1 which presents a smooth shape.

In Fig. 6.33, results of the trajectory tracking during the full trial in scenario 1 is shown. Again, the resulting vehicle behaviour when following the planned trajectories is similar to those obtained when the occupancy grid is not used for motion planning purposes. Nevertheless, it is observed a higher amount of planning requests are performed in this case (19, in contrast to 13 in the case of Sect. 6.4.5.1), as can be seen on Fig. 6.33D. This is due to the shorter distance ahead used in this case in order to obtain a reasonable grid size. Note that the higher the grid size, the longer the grid computing time. Moreover, as in the case in Sect. 6.4.5.1, all planning requests are performed with mode *1* except the first one, which is performed with mode *0*, as expected since there is no obstacles in this scenario.

Finally, Fig. 6.33C shows the reference speed and measured vehicle speed during this trial.

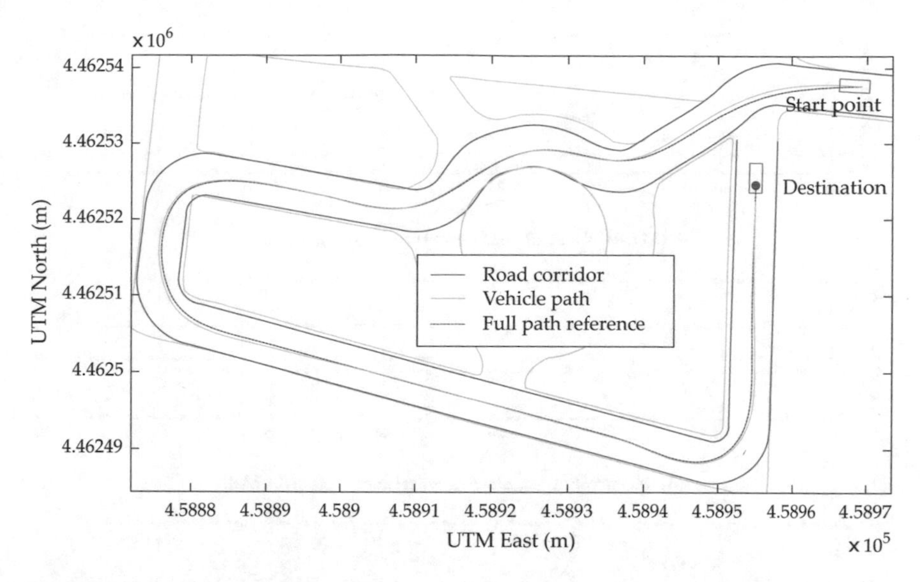

Fig. 6.32 Final reference path and vehicle path in the scenario 1 using the occupancy grid

Regarding the planning time, the mean of the processing time per planning request is 112.99 ms with a standard deviation of 64.10 ms. Although this results shows that the planning time last approximately twice as long as when the occupancy grid is not used, it is still reasonable.

To analyze the resulting behaviour of the vehicle in terms of occupant comfort, Fig. 6.34 shows a density plot of the real longitudinal and lateral accelerations to measured along the trial. As in the similar case in Sect. 6.4.5.1, this figure depicts that most of the acceleration measurements fall within the dashed white rectangle that represent the stated acceleration limits.

Figure 6.35A, B show the computation time and the number of grids of grids generated during this experiment, respectively.

The histogram in Fig. 6.35B depicts the distribution of the number of computed grids during the test (60 s approximately), according to their sizes. As can be seen, the size of most grids is between 20,000 and 50,000 cells in this case. Figure 6.35A shows the computation time of the grids with respect to their size of all the grids computed. In this case, the computation times of most grids concentrates between 60–90 ms, what is a reasonable value.

In conclusion, this experiment showed that the proposed motion planning algorithm using the occupancy grid as input is able to generate a number of valid candidates and select the optimum candidate in a few milliseconds, even in sharp areas where consecutive curves must be overcome by the vehicle. The main difference with regard to the same experiment carried our without the occupancy grid relies on the computation time, obtaining a worse performance in this sense. However, this

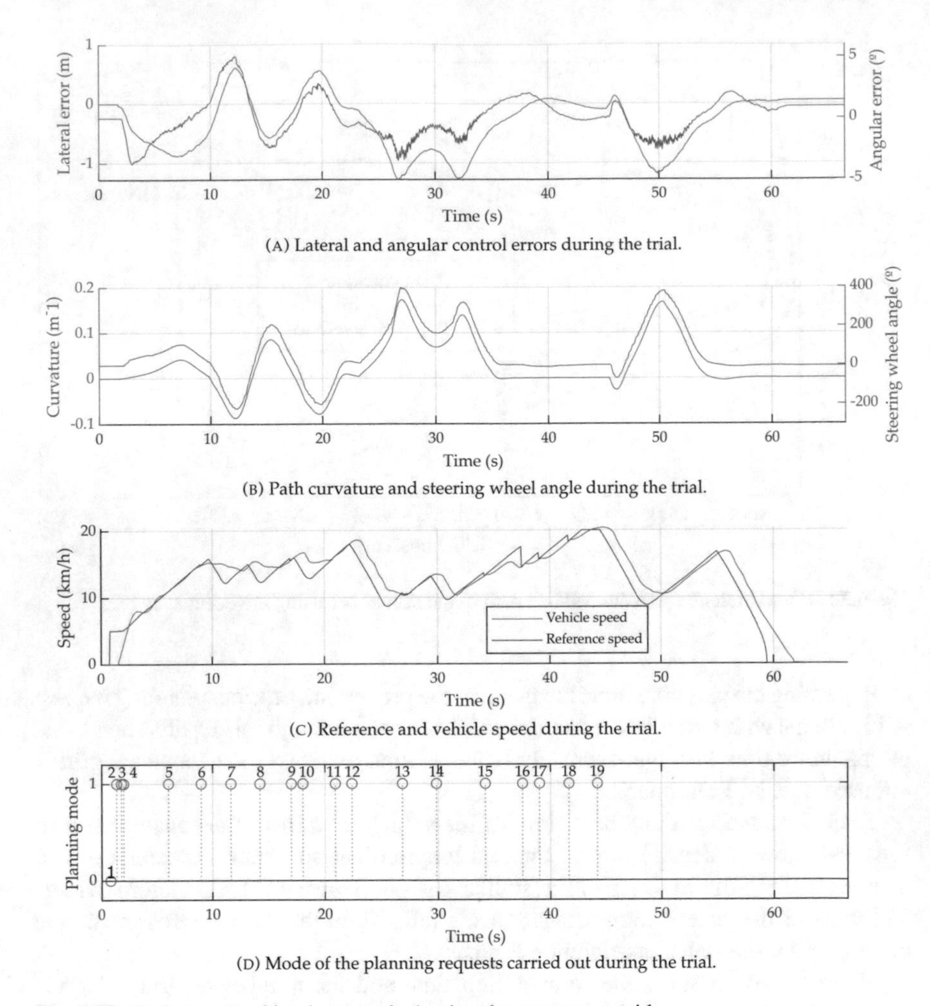

Fig. 6.33 Trajectory tracking in scenario 1 using the occupancy grid

method consider the localization uncertainty as it has been previously propagated over the occupancy grid.

6.5.2.2 Scenario 2: Static Obstacles Avoidance Using the Occupancy Grid

This scenario is similar to the one presented in Sect. 6.4.5.1. It includes static obstacles located at different places of the route that the vehicle is following to reach the destination point. Thus, this scenario includes a greater complexity with respect to the previous one.

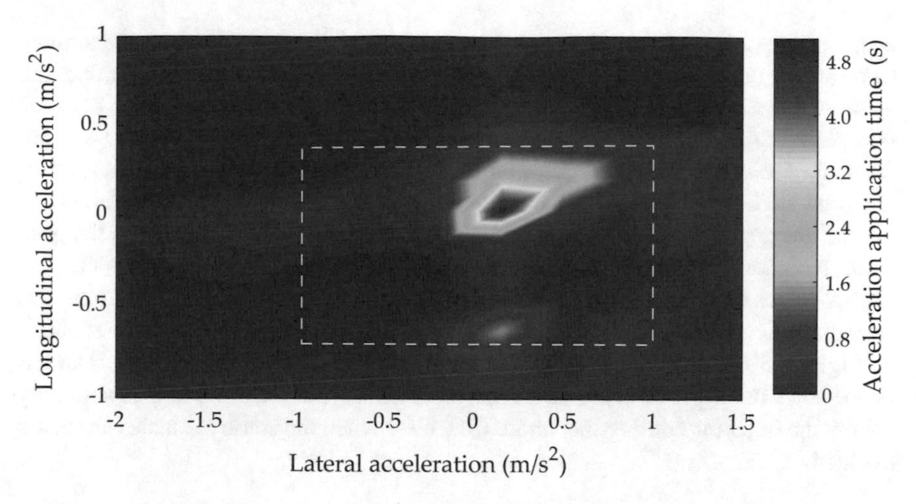

Fig. 6.34 Density graph of measured acceleration in the vehicle during the trial in scenario 1 using the occupancy grid

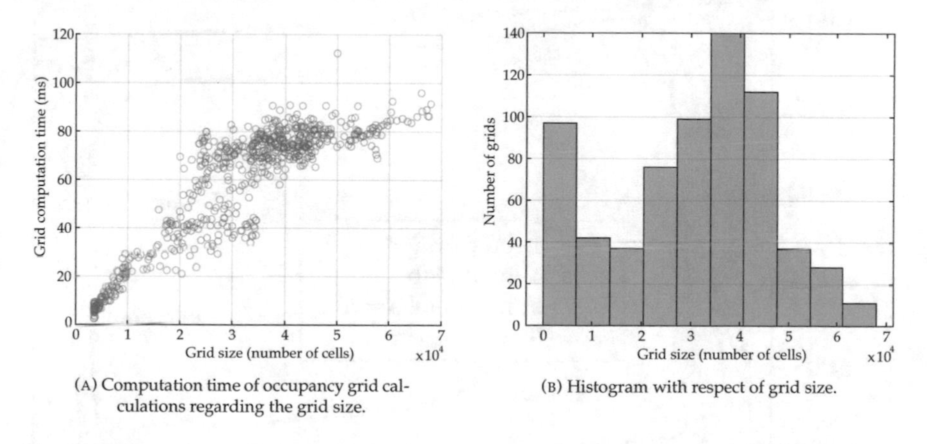

(A) Computation time of occupancy grid calculations regarding the grid size.

(B) Histogram with respect of grid size.

Fig. 6.35 Occupancy grid results in scenario 1

Figure 6.36 shows the concatenation of final paths together with the real path of the vehicle. In this case, just like in the similar scenario without using the occupancy grid (Sect. 6.4.5.1), the trajectory is quickly corrected to avoid the obstacles satisfactorily even with noisy perception information. In addition, Fig. 6.37 shows two consecutive screenshots of the 3D visualization while the vehicle was avoiding the obstacles and Fig. 6.38 shows two front vehicle pictures at similar instants of the trial, where static obstacles (two cardboard boxes) can been seen.

Regarding computing time, the average planning time for the whole trial was 193.18 ms with a standard deviation of 144.58 ms for the whole experiment. In comparison with the previous case, the mean planning computation time has been increased due to the static obstacles avoidance requests. Note that depending on the shape of the ahead section of the road corridor that fall within the grid, the amount of reference points used to generate path candidates vary along the trial. Taking into account that the occupancy grid has been parametrized to achieve a balance between computing time, ahead distance and planning space exploration, typically a lower number of reference points are used and consequently a lower amount of candidates are generated. However, when a planning request to avoid a static obstacle is carried out, all the generated reference points (as shown in Fig. 6.22) falls within the grid. Thus, a higher number of candidates are evaluated in these type of requests, causing higher values of mean and standard deviation values. This effect could be mitigated by distributing a low number of reference points when avoiding a static obstacle.

Figure 6.39 depicts the same information about the trajectory tracking than has been shown in the previous cases. As can be observed, the vehicle is able to smoothly follow the trajectory during the whole trial, even when the static obstacles are being avoided.

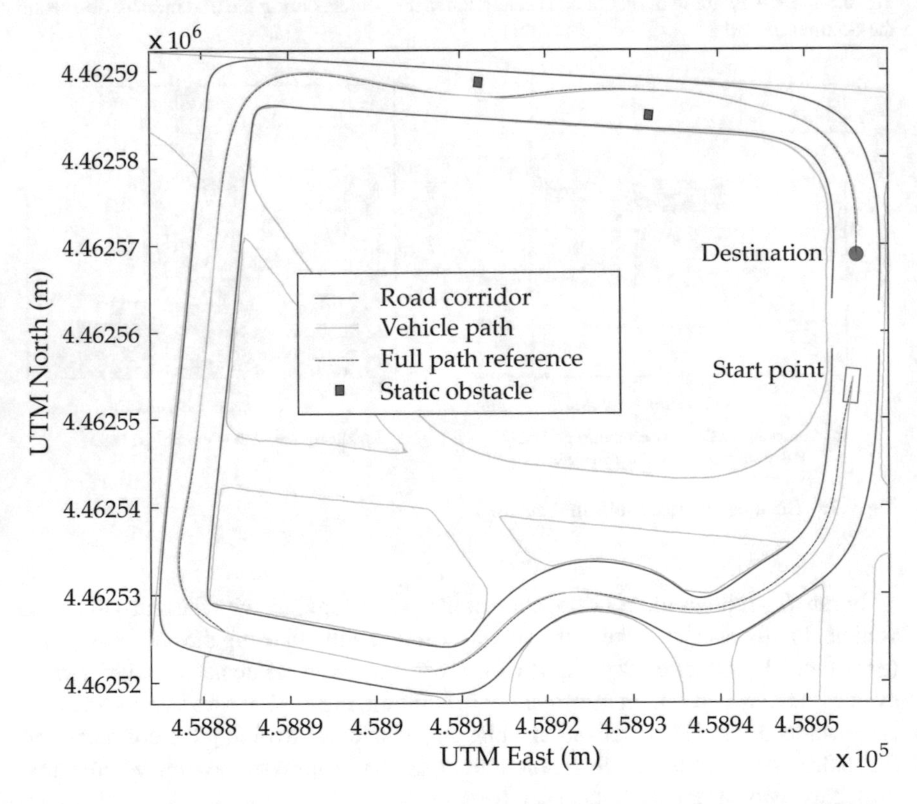

Fig. 6.36 Resulting paths in scenario 2 using the occupancy grid

(A) Avoiding first static obstacle.

(B) Avoiding second static obstacle.

Fig. 6.37 3D visualization screenshots while avoiding static obstacles

(A) Avoiding first static obstacle.

(B) Avoiding second static obstacle.

Fig. 6.38 Pictures of the frontal vehicle view while avoiding static obstacles

In order to analyse the comfort inside the vehicle during the test, Fig. 6.40 shows a density map to represent the measured longitudinal and lateral accelerations. Figure 6.40 shows how most of the measured acceleration values fall within the limits (marked with a dashed white rectangle) established in the planning. However, some values are outside mainly due to the joint effect of vibrations induced by road imperfections and road and vehicle pitching and rolling.

Finally, Fig. 6.41A, B show the computation time and the number of grids of grids generated during this experiment, respectively. In this case, the computation times of the grids concentrates in around 75 ms for most of the grids computed during the test, whose sizes are mostly between 35,000 and 55,000 cells. Although this scenario includes static obstacles, these results are similar to those obtained in the previous experiment. Note that the distribution of the grids according to their size is different from the previous case. This is caused by the different route followed in this scenario.

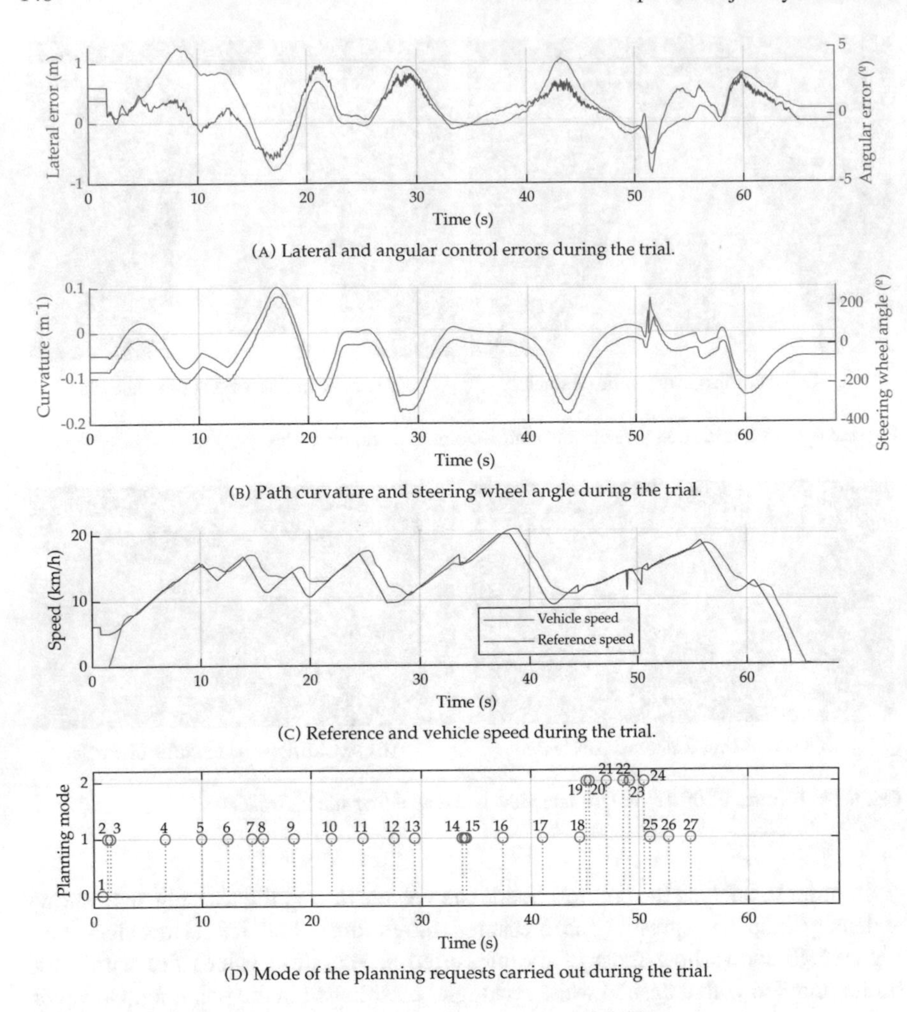

(A) Lateral and angular control errors during the trial.

(B) Path curvature and steering wheel angle during the trial.

(C) Reference and vehicle speed during the trial.

(D) Mode of the planning requests carried out during the trial.

Fig. 6.39 Trajectory tracking in scenario 2 using the occupancy grid

6.5.2.3 Scenario 3: Long Urban Route Using the Occupancy Grid

Figure 6.42 shows the road corridor used together with the full path followed by the vehicle until the destination point is reached. Moreover, Fig. 6.43 shows a screen shot of the 3D visualization of one occupancy grid and path computed (left) and a front picture (right) during a real trial.

The performance in terms of path smoothness of the paths generated using the occupancy grid approach is similar than in the case of the general approach as the cost function is quiet similar in both cases. However, it can be noticed how the path obtained through the occupancy grid approach tends to be more centred in the lane

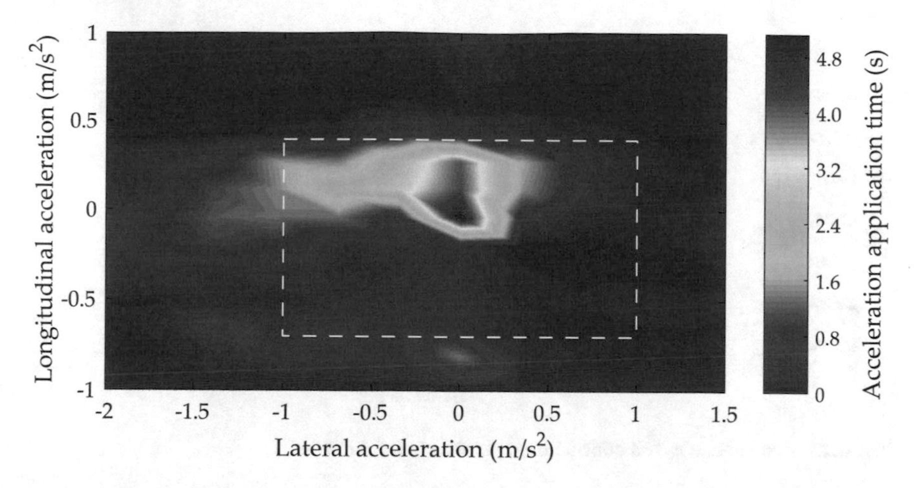

Fig. 6.40 Density graph of measured acceleration in the vehicle during the trial in scenario 2 using the occupancy grid

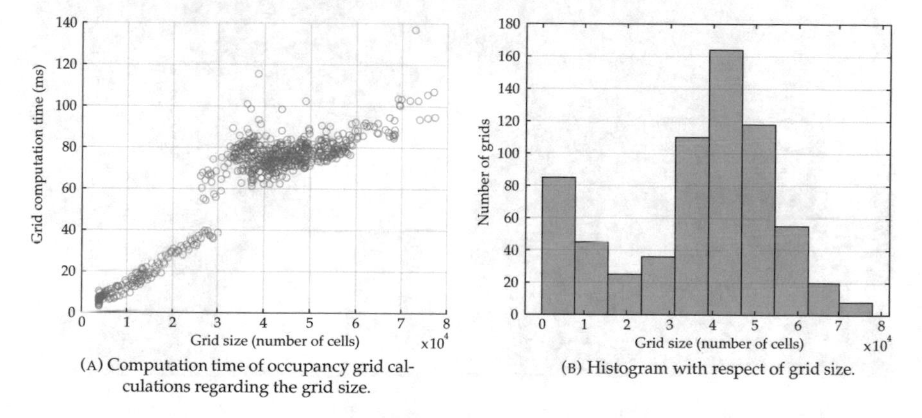

(A) Computation time of occupancy grid calculations regarding the grid size.

(B) Histogram with respect of grid size.

Fig. 6.41 Occupancy grid results in scenario 2

since the occupancy probability is typically lower in the centre of the lane than in the vicinity of the edges.

Regarding the planning computation time, when the occupancy grid-based planning approach is used, it is expected a different performance with respect to the general approach since the computationally expensive process of collision checking with static objects and the "inside corridor" verification is replaced by the probability thresholding of Eq. (6.37). However, the computation time of the uncertainty propagation along the grid must be considered. The mean planning time in this test was 143.51 ms with a standard deviation of 130.08 ms.

The resulting trajectory could be easily followed by the vehicle. Figure 6.44 shows relevant tracking variables logged during the test. It can be noticed the continuity of

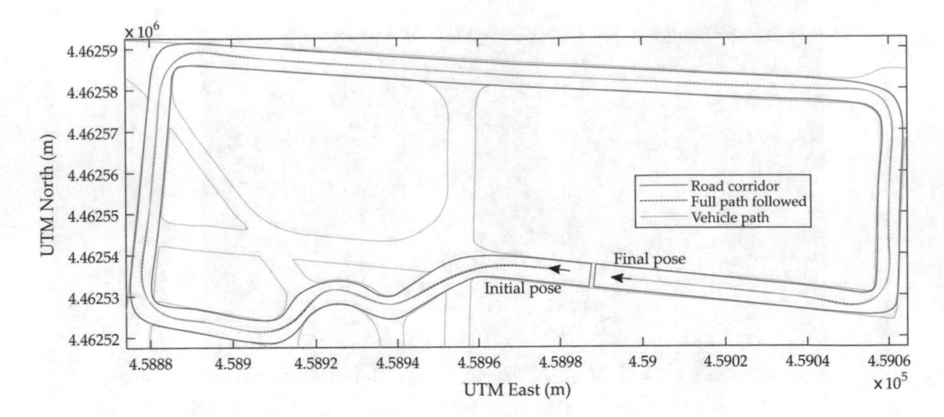

Fig. 6.42 Road corridor, full concatenated planned and vehicle paths

Fig. 6.43 Screen shot of the 3D visualization during the trial

both lateral (lateral and angular error, see Fig. 6.44A; and path curvature and steering angle, see Fig. 6.44B) and longitudinal (reference and vehicle speed, see Fig. 6.44C) control variables.

Furthermore, Fig. 6.45A shows the computation time of the grids with respect to their size of all the grids computed during the test (110 s approximately). Note that the computation times of the grids concentrates in around 70 ms, what is a reasonable value. Besides this scatter plot, the histogram in Fig. 6.45B depicts the distribution of the amount of computed grids according to their sizes. As can be seen, the distribution of the number of grids is sightly different to the previous scenarios since the route followed by the vehicle is different. In this case, the size of most grids is between 20,000 and 50,000 cells.

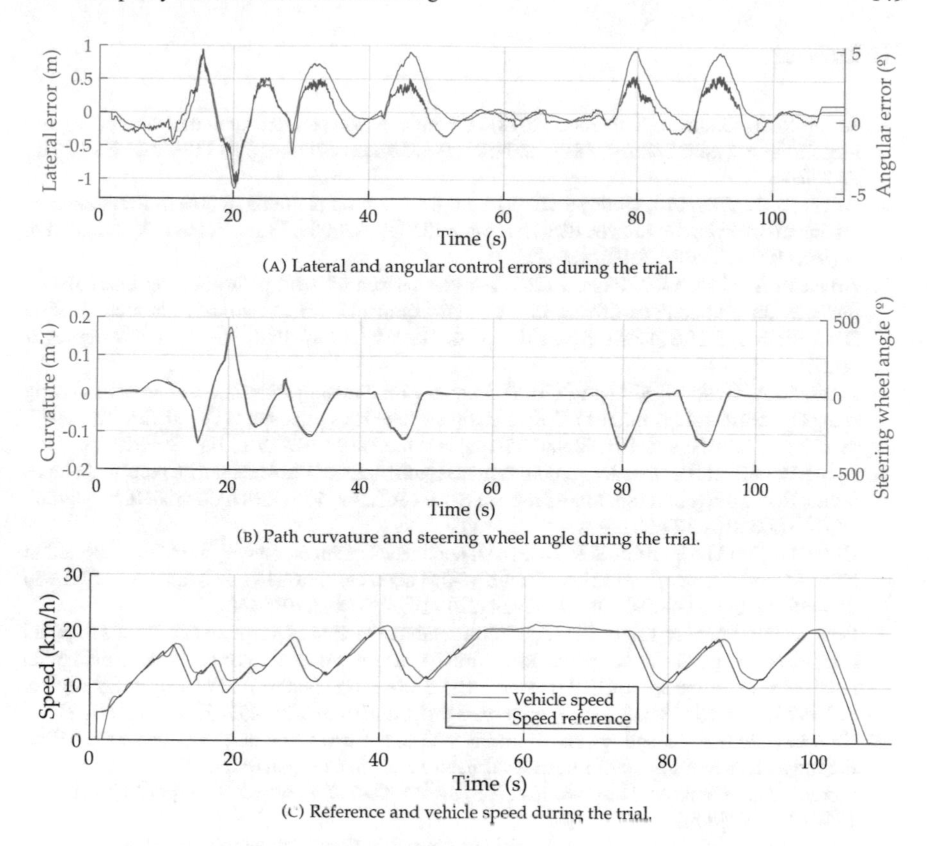

(A) Lateral and angular control errors during the trial.

(B) Path curvature and steering wheel angle during the trial.

(C) Reference and vehicle speed during the trial.

Fig. 6.44 Trajectory tracking information in scenario 3

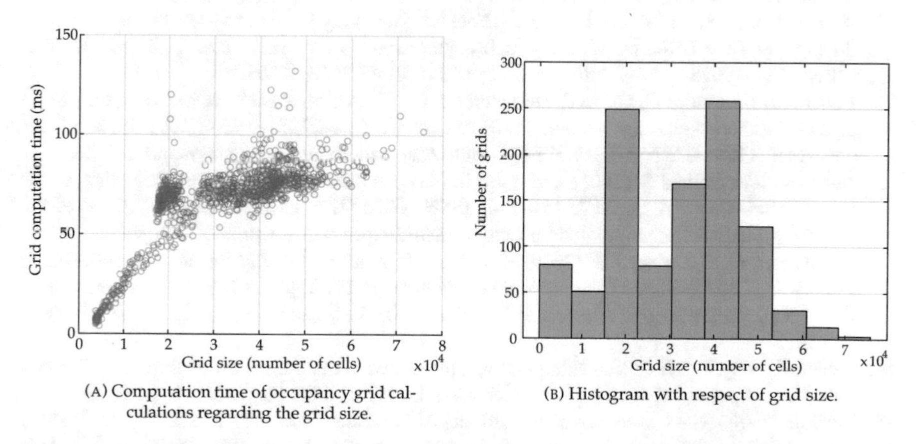

(A) Computation time of occupancy grid calculations regarding the grid size.

(B) Histogram with respect of grid size.

Fig. 6.45 Occupancy grid results in scenario 3

References

1. Artuñedo A, Godoy J, Villagra J (2018) A primitive comparison for traffic-free path planning. IEEE Access 6:28801–28817. ISSN: 2169-3536. https://doi.org/10.1109/ACCESS.2018.2839884
2. Artuñedo A, Villagra J, Godoy J (2019) Real-time motion planning approach for automated driving in urban environments. IEEE Access 7:180039–180053. ISSN: 2169-3536. https://doi.org/10.1109/ACCESS.2019.2959432
3. Artuñedo A, Godoy J, Villagra J (2017) A comparison of local path-planning interpolation methods for autonomous driving in urban environments. In: Industriales research meeting 2017. Madrid: ETSII, UPM, Apr. 2017, p 147. ISBN: 978-84-16397-58-7. http://oa.upm.es/46090/
4. Artuñedo A, Godoy J, Villagra J (2017) Smooth path planning for urban autonomous driving using OpenStreetMaps In: 2017 IEEE intelligent vehicles symposium (IV). IEEE, June 2017, pp 837–842. ISBN: 978-1-5090-4804-5. https://doi.org/10.1109/IVS2017.7995820
5. Byrd RH, Gilbert JC, Nocedal J (2000) A trust region method based on interior point techniques for nonlinear programming. Math Program Ser B 89.1:149–185. ISSN: 00255610. https://doi.org/10.1007/s101070000189
6. Digabel S (2011) Algorithm 909: NOMAD: nonlinear Optimization with the MADS algorithm NOMAD: nonlinear optimization with the MADS algorithm. ACM Trans Math Softw (TOMS) 4:44–15. ISBN: 0098-3500. https://doi.org/10.1145/1916461.1916468
7. Douglas DH, Peucker TK (2011) Algorithms for the reduction of the number of points required to represent a digitized line or its caricature. In: Classics in cartography: reflections on influential articles from Cartographica 10.2 (Dec. 2011), pp 15–28. ISSN: 0317-7173. https://doi.org/10.1002/9780470669488.ch2. http://utpjournals.press/doi/10.3138/FM57-6770-U75U-7727
8. Farin G (2002) 8—B-spline curves. In: Farin G (ed) Curves and surfaces for CAGD (Fifth Edition). Fifth Edition. The Morgan Kaufmann Series in Computer Graphics. San Francisco: Morgan Kaufmann, pp 119–146. ISBN: 978-1-55860-737-8. https://doi.org/10.1016/B978-155860737-8/50008-9
9. Ganesh M (2008) Basics of computer aided geometric design: an algorithmic approach. I.K. International. ISBN: 9788189866761
10. Gu T, Snider J, Dolan JM, Lee J-W (2013) Focused Trajectory Planning for autonomous on-road driving. In: 2013 IEEE intelligent vehicles symposium (IV). IEEE, June 2013, pp 547–552. ISBN: 978-1-4673-2755-8. https://doi.org/10.1109/IVS.2013.6629524
11. Haber RE, Beruvides G, Quiza R, Hernandez A (2017) A simple multi-objective optimization based on the cross-entropy method. IEEE Access 5:22272–22281. ISSN: 2169-3536. https://doi.org/10.1109/ACCESS.2017.2764047. http://ieeexplore.ieee.org/document/8070310/
12. Hormann K, Agathos A (2001) The point in polygon problem for arbitrary polygons. Comput Geome: Theor Appl 20(3):131–144. ISSN: 09257721. https://doi.org/10.1016/S0925-7721(01)00012-8. https://www.inf.usi.ch/hormann/papers/Hormann.2001.TPI.pdf
13. Lau B, Sprunk C, Burgard W (2009) Kinodynamic motion planning for mobile robots using splines. In: 2009 IEEE/RSJ international conference on intelligent robots and systems. IEEE, Oct. 2009, pp 2427–2433. ISBN: 978-1-4244-3803-7. https://doi.org/10.1109/IROS.2009.5354805
14. Levien R, Séquin CH (2009) Interpolating splines: which is the fairest of them all? Comput Aided Des Appl 6(1):91–102. ISSN: 16864360. https://doi.org/10.3722/cadaps.2009.91-102
15. Moré JJ (1978) The Levenberg-Marquardt algorithm: implementation and theory. In: Watson GA (ed) Numerical analysis. Berlin, Heidelberg: Springer Berlin Heidelberg, pp 105–116. ISBN: 978-3-540-35972-2

16. Opheim H (1981) Smoothing a digitized curve by data reduction methods. In: Encarnacao JL (ed) Proceedings of the international conference. The Eurographics Association, p 127. ISBN: 0444862846. https://doi.org/10.2312/eg.19811012. http://diglib.eg.org/EG/DL/Conf/EG81/papers/EUROGRAPHICS
17. Zhang Y, Chen H, Waslander SL, Gong J, Xiong G, Yang T, Liu K (2018) Hybrid trajectory planning for autonomous driving in highly constrained environments. IEEE Access 6:32800–32819. ISSN: 2169-3536. https://doi.org/10.1109/ACCESS2018.2845448. https://ieeexplore.ieee.org/document/8375948/
18. Ziegler J, Stiller C (2010) Fast collision checking for intelligent vehicle motion planning. In: 2010 IEEE intelligent vehicles symposium. IEEE, June 2010, pp 518-522. ISBN:978-1-4244-7866-8. https://doi.org/10.1109/IVS.2010.5547976

Chapter 7
Integration and Demonstrations

7.1 Introduction

Once all decision-making architecture modules have been presented in Chaps. 4–6, this chapter describes how they are integrated in the architecture presented in Chap. 3. For that purpose, the detailed description of the experimental platform used to evaluate the proposed architecture in a real environment is firstly provided in Sect. 7.2. This section focuses in both hardware and software aspects of the vehicle components, providing an in-depth description of sensor and actuation systems as well as a detailed insight of the software architecture.

Besides the specific validation of the architecture modules that has been presented in Chaps. 4–6, several live demonstrations, detailed in Sect. 7.3, were carried out to show the set of capacities of the proposed architecture.

7.2 Experimental Platform

This section introduces the experimental platform used to test and validate the proposed decision-making algorithms in a real environment.

7.2.1 Experimental Platform Components

With the goal of demonstrating the system reliability and its possible applications, the architecture has been implemented in one of the cars of the AUTOPIA Program fleet. Nevertheless, it is applicable to any test car as the implementation of each module has been developed as general as possible, providing only those customizations that are only necessary for its application to this specific car.

A. Artuñedo, *Decision-making Strategies for Automated Driving in Urban Environments*, Springer Theses, https://doi.org/10.1007/978-3-030-45905-5_7

The car is a conventional Citroën DS3 which includes hardware modifications for the automated control of throttle, gearbox, brake and steering systems. Moreover, a set of proprioceptive and exteroceptive sensors have also been installed. As on-board unit, the test car has one standard PC installed in the trunk. This computer runs a linux-based OS on a machine with an Intel i7-3610QE processor, 8 GB of RAM memory and a solid-state drive.

The first step to control any system is the data acquisition, and for a self-driving car this is not different. When a person is driving a car on the road, he manages information about his own vehicle, the road and the nearest cars at the same time. In this section, the on-board sensors considered in the implemented perception stage are presented. Nevertheless, note that the architecture is open to include any additional source of information.

The following subsections introduce the installed sensors, controllers, actuators as well as the software architecture.

7.2.1.1 Proprioceptive Sensors

CAN Bus

Nowadays, cars have a high number of on-board sensors to capture data about the vehicle state. This information is used at the same time by several modules with different purposes—e.g. the information about the wheels speed is used for showing the driver the current speed of the vehicle, but it is also used by the anti-lock braking system (ABS) and the electronic stability program (ESP). To share all this data, cars have an on-board network, usually under the CAN protocol, to interconnect all the sensors and modules.

Having so much data flowing through the vehicle network, the idea of incorporating this information to the control architecture for automated driving is straightforward. To achieve this goal, a PCI CAN adapter is used to connect the on-board PC to the vehicle network as another on-board module. However, the writing capability of the adapter has been removed for safety reasons, so this connection does not interfere with the performance of any ECU in the car. Once the computer is part of the car network a program is able to receive and decode all the data in a similar way it was made with previous sensors.

Due to the different rates and nature of the data coming from the CAN BUS, the CAN reader module has been deployed with a set of different output channels:

- `CAN_CAR_GEARBOX`: Contains gearbox state information such as current gear change mode, current engaged gear or current requested gear.
- `CAN_CAR_CONSUMPTION`: Contains engine consumption data.
- `CAN_CAR_STEERING`: Include steering wheel position and speed, steering column torque, etc.
- `CAN_CAR_BRAKE_ABS`: Contains the braking pressure, yaw rate and lateral acceleration.

- CAN_CAR_CABIN: Includes the activation state of brake pedal (pressed or not pressed), ABS activation and vehicle ignition state.
- CAN_CAR_ENGINE: Contains the engine speed, real torque and requested torque.
- CAN_CAR_SPEED: Contains vehicle longitudinal speed and acceleration, and the speed of each wheel separately.

Inertial measurement unit

Although a DGPS equipment could be considered precise enough to estimate the vehicle position, there are some scenarios where the GPS measurements are affected by interference in the satellite signals (e.g. buildings, trees or clouds) reducing the accuracy of the measurement [9]. Moreover, the vehicle could be in an area where there is not a clear sky-view such as tunnels or under bridges, making impossible to get a reliable position just using a GPS unit. A common solution to this problem is to incorporate inertial sensors to the architecture, meaning that even when there is not GPS availability, the position could be estimated from previous ones through Dead Reckoning techniques [6].

In our case, a Crossbow VG440 unit has been installed near the centre of gravity of the vehicle. This inertial measurement unit (IMU) provides information about the angular rate and acceleration of the vehicle in three axes with a sampling rate up to 100 Hz. The data is transmitted by serial port to the on-board unit, where a dedicated program decodes it and transmits it as a LCM message. A serial-connection between the GPS and the IMU allows synchronizing both sampling and reference time with the internal clock of the GPS unit.

7.2.1.2 Exteroceptive Sensors

GPS

In this architecture, the GPS represents the backbone of the localization system for two reasons: on the one hand, by using a differential GPS equipment (DGPS) it is possible to get the vehicle position with a very high accuracy; and on the other hand, the GPS receivers have a high-accuracy internal clock that can be used as control clock reference for the entire system.

For the implementation we have selected the model Trimble BD960. This device is able to work with a rate up to 20 Hz. The prior architecture was designed to work with 10 Hz GPS data. However, it has been configured to provide localization information at 20 Hz in the new architecture since higher accuracy can be achieved by the developed state estimator.

The receiver offers centimeter-level accuracy based on carrier phase RTK. It is able to use the differential GPS correction obtained through the WAAS/EGNOS network or from nearby reference stations, increasing the position accuracy up to 2 cm. As output, this equipment provides several data frames under the NMEA0183 protocol [17] through both serial and socket interfaces.

Fig. 7.1 DS3: the test platform

In order to process the data and make them available to all the system, a decoder program reads three NMEA frames through one of the available interfaces and publishes it in a dedicated LCM channel. From the GGA, VTG and GST messages it is possible to extract information about the vehicle: position (in geographic coordinates), speed, orientation and altitude; resolution mode of the GPS: fix quality, error ellipsoid and standard deviation. Before building the LCM message and transmitting it over the channel, the program calculates the position in UTM coordinates.

LiDAR

The vehicle is equipped with a LiDAR (Ibeo Lux 4l) sensor on the front (see Fig. 7.1). This sensor is used to detect the surroundings of the vehicle and/or the objects located within the field of view. To that end, it scans the surroundings with several rotating laser beams, receives the echoes with a photo diode receiver, processes the data by means of a time of flight calculation and issues the processed data via the interfaces Ethernet and/or CAN.

The sensor characteristics are shown in Table 7.1.

The object data include information on a high level. Instead of raw data, a set of objects and their properties (such as size, position, absolute/relative speed and object type) is provided. The sensor has been configured to issue the object data at 12.5 Hz.

Table 7.1 LiDAR sensor specifications

Specifications	Ibeo Lux (4 layers)
Horizontal field	85° (35° to −50°)
Horizontal angular resolution	0.125°
Vertical field	3.2°
Vertical angular resolution	0.8°
Range	200 m
Update frequency	12.5/ 25.0/ 50.0 Hz

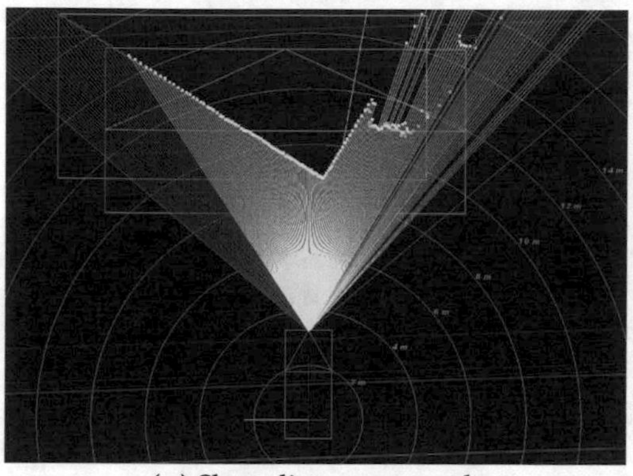

(A) Short distance example

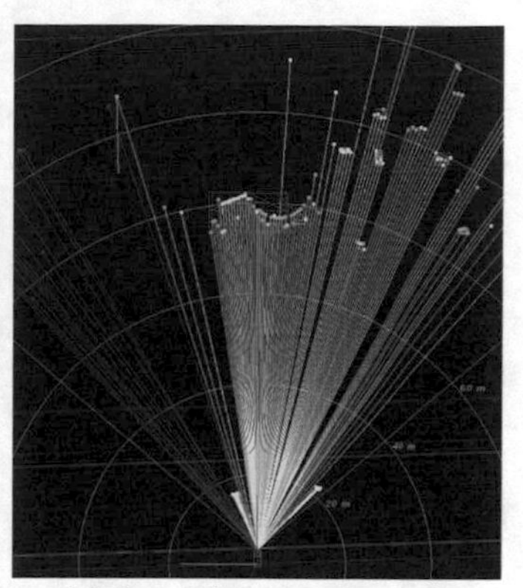

(B) Long distance example

Fig. 7.2 2D representation of LiDAR output data

These data have been used to validate the motion planning algorithms proposed in this thesis.

An example of the data provided by the Ibeo Lux 4l is shown in Fig. 7.2, where raw data is represented by straight segments from the sensor to the impact point, and are coloured according to the layer they belong. Moreover, the detected objects are bounded by rectangles.

Stereoscopic camera

Vision sensors are usually equipped in automated vehicles to provide useful information about the surroundings that could assist its navigation. In this sense, our experimental platform incorporates a stereoscopic sensor that can be used to compute depth information from two synchronized and perfectly correlated images.

The camera is located inside the car behind the rear-view mirror at a height measured from the ground of 1290 mm. In order to perceive as much relevant information as possible assuming right lane driving, the camera is placed with a yaw angle of 2° and a pitch angle of 5.2° with respect to the car reference system. In this way, the field of view covers part of the left lane.

The camera sensor specifications are shown in Table 7.2.

Table 7.2 Stereo camera sensor specifications

Specification	Bumblebee2
Resolution	1032 × 776
Frame rate	20 FPS
Megapixels	0.8 MP
Sensor name	Sony ICX204
Sensor type	CCD
Sensor format	1/3"
Pixel size	4.65 μm
Focal length	3.8 mm, 65° HFOV
Aperture	f/2.0
Interface	FireWire 1394a
Dimensions	157 × 36 × 47.4 mm

The stereoscopic camera is used provide visual information to the vision-based road corridor adaptation algorithm included in the proposed architecture.

7.2.1.3 General Inputs and Outputs

This section covers a description of several hardware components that can be used at the same time for different software modules in different stages. The section is divided in two groups of components: I/O modules and Communication modules.

Input and output modules

The previous section includes some of the sensors that are fundamental to acquire the needed data to control the vehicle automatically. However, there are other variables in a car that should be sampled in order to increase the safety of the entire system and improve the experience of a human driver. Most of these values can be measured directly from analogue or digital signals in the car wiring. For example, it is possible to determine whether the driver is braking the car by reading the state of the brake pedal. Moreover, the same wiring could be used to turn on the braking lights when the autonomous control system is activated. For these reasons, the installation of digital/analogue I/O modules is fairly convenient.

For the implementation of AUTOPIA fleet, we have selected two different I/O modules. The first one is a module that can be controlled with a CAN bus. This unit has 2 analog inputs, 2 analog outputs, 4 digital inputs and 8 digital outputs. Each digital output is associated to the activation of one relay, reducing the limitation to control signals with different reference values. The second unit is a PCI analogue output card with a 16-bit DAC. This card allows using analogue values requiring a higher resolution since the first unit has a 0.1 V accuracy.

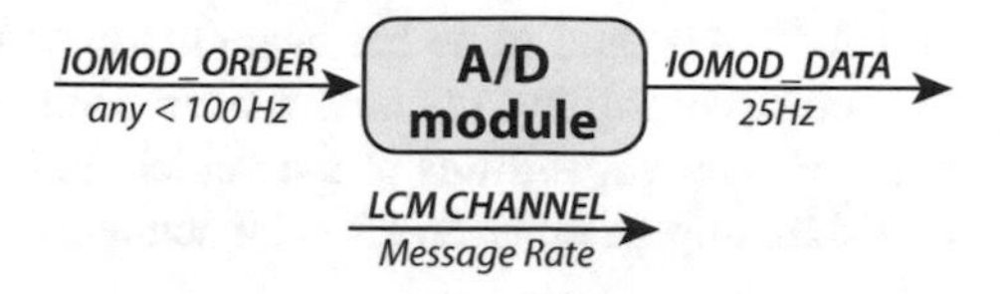

Fig. 7.3 LCM input and output channels for one of the I/O modules implemented

To allow other software modules in the architecture to use these units, each of them is managed by an adequate control program. This software has two main goals: (i) it reads all the module inputs and publishes this information in one LCM channel at 25 Hz. (ii) the program subscribes to a dedicated LCM channel where the output commands are published by the external programs that need to use the unit. In each command message, a priority field has been included. This allows the unit control program to manage the orders correctly when several modules are sending commands at the same time. An example of this program in one of the implemented software modules is schematized in Fig. 7.3.

Communications

A human driver uses the information about his own car, as well as information about other vehicles and the environment. A common approach to make this possible with an autonomous car is the implementation of communications systems that allow the vehicles to exchange data among them and with the road infrastructure [23, 24]. Through this communications links, the on-board control programs are able to get information about the localization of the other cars as well as the relevant information about the road as traffic and weather conditions.

In order to provide our architecture with the capability for information exchange with other vehicles and the infrastructure, two hardware modules have been integrated. The first module is a PCMCIA card that works under the WiFi protocol. This card permits the interaction between the car and a central system that manages the information coming from all the vehicles in the nearby area. Details of this implementation are widely described in [12, 13]. The second system is the communication box developed for the Grand Cooperative Driving Challenge, which is based on the IEEE 802.11p [25]. This system has been developed for both V2V and V2I communications.

From the point of view of the other modules in the architecture, both communications units are totally transparent and can be used indistinctly as UDP sockets from each program. However, to preserve a certain order in the communication system and reduce the amount of data transmitted over the links, it is recommended that only the main control program will implement the communications part.

7.2.1.4 Actuation

We will describe now the actuation stage before considering the development of the decision stage. As with a human drivers, the on-board computer must be able to control, at least, the principal actuators of the vehicle: throttle, brake and steering wheel. To this end, some hardware modifications must be applied over each actuator before controlling them. However, each one of these modifications depends on the particular capabilities of the vehicle to be controlled. This section presents the hardware implementation for each actuator in our vehicle.

Throttle

Under normal conditions, the throttle in our vehicle is controlled by an electronic pedal that transduces the pressure applied by the driver as two analog signals (one of them twice the other) and sends them to the engine controller. In order to emulate the behaviour of the pedal, a low-level program has been developed. This software is in charge of reading all the throttle commands sent by the control program through a LCM channel with the final goal of determining and sending the corresponding orders to the I/O analogue card to generate the adequate signals to control the throttle.

For the development of the program, two important considerations have been made: First, throttle command values must be normalized, meaning that only values between 0—no throttle—and 1—maximum throttle—are accepted. Values out of this range are approximated to the nearest limit. The second consideration is related to safety. After a throttle command is processed and sent to the analogue card, the program waits up to 250 ms for a new command to arrive. If no command is received within this time then the program disables the throttle. This prevents the throttle from keeping the last value in case the control program crashes or it is ended by the user.

A switch installed in the board allows the driver to commute between manual and automatic throttle. The state of the switch is associated to two relays that toggle the throttle connections between the electronic pedal and the analog card output.

Brake

The brake system plays a key role in the safety of the vehicle. Thus, its automation is as important as tricky since it must not interfere with the standard vehicle brake. As in other vehicles of AUTOPIA Program [19], an electric motor was coupled to the brake pedal allowing the computer to brake the car (see Fig. 7.4).

As it was made for throttle control, a low-level program has been developed to control the position of the brake pedal. The software subscribes to a reserved channel for brake commands and processes all the received messages. For each command the program calculates the corresponding value of the valves control signals and sends the order to the low-level controller. This program implements normalized brake commands between 0 and 1, where 0 means no brake and 1 the maximum brake pressure. As for throttle, a safety timer has been also implemented, disabling the automatic brake when an order has not been received for 250 ms.

The automatic braking mechanism allows to automatically press the brake pedal as a human driver would do, allowing also the occupant to press the pedal if needed. In addition, a pressure sensor has been installed under the pedal pad i.e. where the driver applies the braking force. A dedicated microcontroller is in charge of detect if the brake pedal is pressed by the occupant in order to override the automatic brake control system if needed.

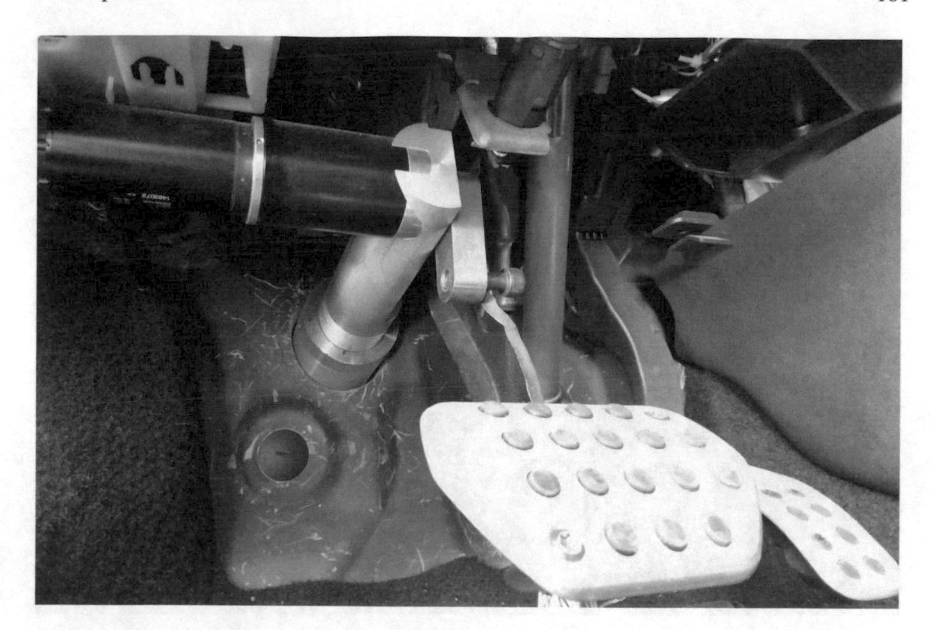

Fig. 7.4 Brake actuator

Steering wheel

The steering wheel is the most delicate actuator to be automated. Due to the reduced available space, the addition of external mechanical systems is difficult and the limitations in performance are not negligible. As in previous works [15], a low-level position control has been developed and implemented. It is composed by an electric motor coupled to the steering bar through a set of gears. The control was performed by a hardware module and an encoder connected to the back of the motor.

In our case, the vehicle has an electric power steering system designed to reduce the driver effort when turning the steering wheel. Although this electric power steering motor is not accessible to be controlled, this system reduces the load of the externally installed motor. Moreover, among the information available over the CAN bus, it is possible to extract the position and turning speed of the steering wheel with a 100 Hz rate. Using this resource we have developed a low level architecture for steering angle control.

An external motor controller was selected due to its high current capacity (up to 60 A) at low voltage levels (12–24 V) is in charge move the steering wheel when the automatic mode is activated (see Fig. 7.5). It has two control interfaces: RS-232 and CAN bus, that can be used to receive control references. To control the steering wheel position, a PID controller has been used.

The developed software subscribes to a dedicated channel where the position reference for the steering is published as a normalized value between -1 (maximum at right) and 1 (maximum at left). Information about steering position and speed is read from the output LCM channel of the software implemented for the CAN bus

Fig. 7.5 Steering wheel actuator

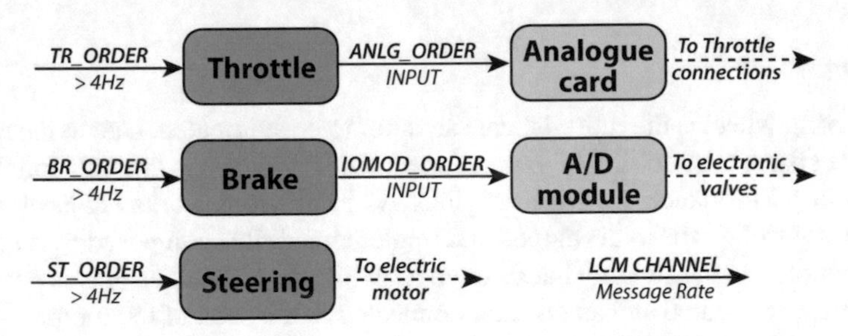

Fig. 7.6 Summary for the actuation stage

of the car (see Sect. 7.2.1.1). The control loop is executed each time that a steering message from the car's CAN bus is received, meaning a control frequency of 100 Hz. As safety measurement, the program leaves the motor in a free wheel mode if a reference command within the 250 ms-range after the last command, so a human driver can retake the control in case of failure. Figure 7.6 shows a summary of the actuation stage modules.

7.2.2 Software Architecture

After defining the general outline of the experimental platform components, the next step is determining how modules will communicate with each other. When a one-block program is running, a memory space is allocated, dynamically or statically, to store all the program variables. As result of this, the different classes and threads of a program can share variables and function calls in an easy way. However, when software is decomposed in several processes, the information exchange among them is not trivial. This problem is well known in computer science and it is studied as interprocess communication (IPC).

Nowadays, several solutions to the IPC problem are available in the commonly used Operating Systems (OS). The most used ones available for UNIX-based OS are listed below:

- Shared and mapped memory: allow process to communicate by reading and writing to a specified memory location.
- Pipes and FIFOs: permit sequential communication among process.
- Sockets: allow communication between different processes even if they are executed in different computers.

The choice of the most adequate IPC technique for an application depends on different factors as: relationship between the processes (related or unrelated), number of processes to communicate with, synchronization among them, writing and reading limitations and number of host computers used [14].

Beyond the listed options, several communications packages have been already developed for different platforms and applications. These packages offer a more developers-friendly option to implement IPC and, in some cases, help to keep a standard among developers regardless of the development platforms or programming languages. One of these alternatives is the Lightweight Communications and Marshalling library (LCM), which is described in detail in the next subsection.

7.2.2.1 LCM as IPC Mechanism for Software Architecture Components

Lightweight Communications and Marshalling (LCM) provides a set of libraries and tools for IPC in real-time systems. It was originally developed and used by the MIT DARPA Urban Challenge Team as message passing system for its vehicle [10]. Its designers have presented LCM as a low-latency, high-throughput solution and it has been compared with similar systems oriented to the robotic field such as ROS [21], presenting significant advantages [8].

LCM is based on a publish-subscribe message passing model using UDP multicast as underlying transport layer [7]. Under this model, processes publish data over a particular channel identified by a unique name and subscribe to those channels required to complete their tasks. Moreover, by using UDP multicast the system becomes highly scalable since the bandwidth required for the transmission of one message is independent of the number of subscribers.

The first benefit of using LCM package is that there is not a set of predefined messages. They are defined and customized by the developers according to their application requirements. To achieve this goal, LCM supplies its own type specification language and a code generation tool that automatically generates a language-specific representation of the message. Supported languages are C, C++, Java, Python, MATLAB and C# [8]. As second advantage, LCM provides each message definition with the marshalling capabilities to encode and decode the data transmitted over the network, instead of using long strings of human-readable messages [7].

In addition to these advantages, LCM provides several tools that are very helpful to develop or inspect a system, namely (i) the logging/replaying tools which allow recording the LCM traffic in a file for future playback or analysis. (ii) A spy tool that analyses, decodes and displays the live LCM traffic. Moreover, the latter tool provides statistics for each channel as message rate, message counter and bandwidth consumption. Both tools have been developed so that their use does not affect the system performance.

7.2.3 *Implementation and Integration of New Software Modules in the Architecture*

The new integrated modules are summarized in the following list:

- Adjustments on the control module.
- Vehicle state estimation module.
- LiDAR objects perception module.
- Camera perception module.
- Human-Machine Interface module.

Figure 7.7 depicts all modules implemented in the architecture. The following subsections describe in detail the implementation of the new modules and functionalities added to the architecture.

7.2.3.1 Local Planner Implementation

The local planner and control modules have been integrated in a multi-threaded C++ application. Taking into account that the generated trajectories are only required by the control module (as depicted in Fig. 7.7), the main motivation to integrate both modules is to avoid the latency in the case that the generated trajectories would be sent through LCM. In this way, the trajectory used by control module is updated as fast as possible after it is generated.

Figure 7.8 shows an UML sequence diagram to show how the manoeuvre planner thread interacts with other architecture components in the case that a new planning request is needed after analysing the perceived objects.

The candidates evaluation stage is the most computationally expensive task of the trajectory generator of the local planner as it has to evaluate the validity and cost of each candidate. With the goal of achieving a motion planner as deterministic

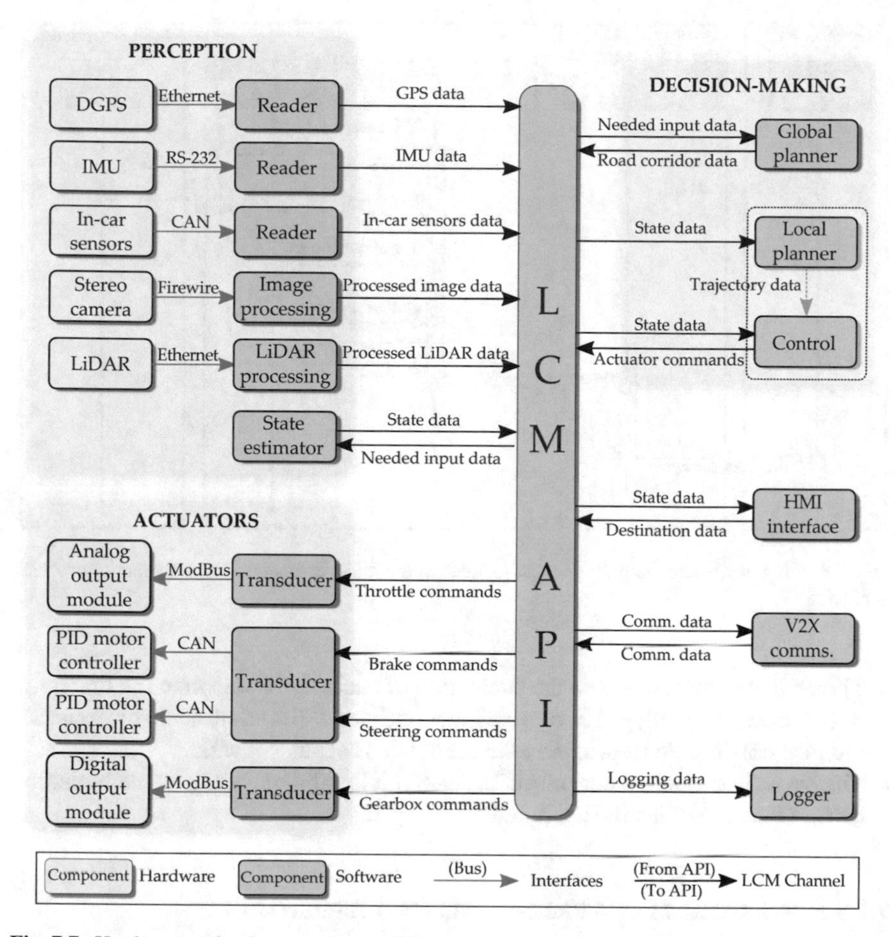

Fig. 7.7 Hardware and software scheme of the architecture

as possible, the quantity of candidates is fixed for every planning request during a mission. In this way the computation time of the trajectory generator is limited when a new trajectory is requested. It should be noted that the computation needed for each candidate is regardless of their length as the discretization of the parametric curve and curvature of each candidate is carried out with the same resolution. As parameters such as number of candidates and discretization resolution plays a key role in the performance of the algorithm, a configuration file is used to set-up these and other parameters.

Besides the limitation of the quantity of the candidates, three evaluation details are added to considerably reduce the computation time:

- The collision checking algorithm is only run if the maximum allowed curvature of the candidate is not exceeded. As a result, the collision checking algorithm is not launched when a candidate is known to be invalid.

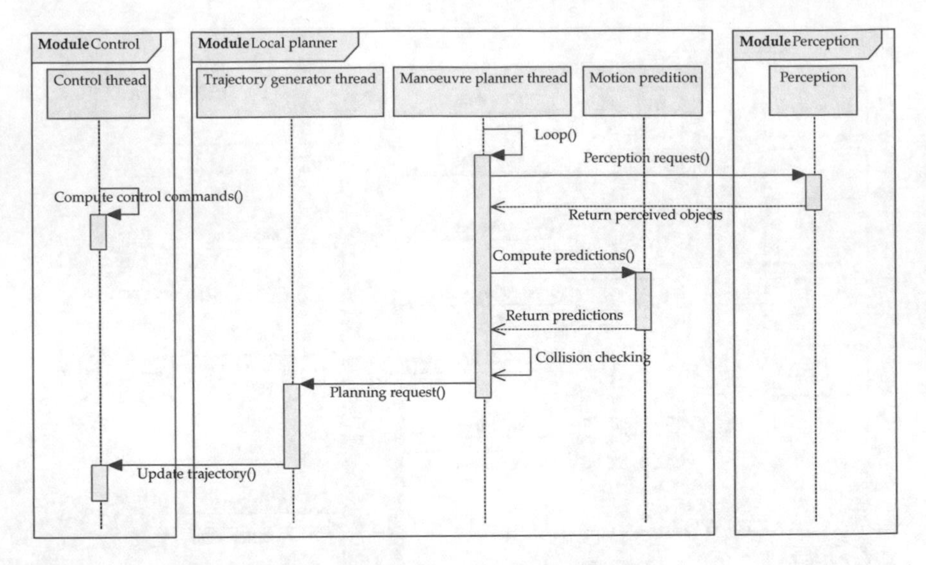

Fig. 7.8 UML sequence diagram of local planner interactions in case that a new planning request is needed

- The collision checking ends the evaluation of each candidate when the first point of the occupancy polygon is detected that collides with and obstacle or is outside the road corridor. In these cases, the candidate is set as not valid.
- The cost of the candidate is only computed if it is valid, reducing the computation time in the case of invalid candidates.

7.2.3.2 Global and Local Planners and HMI Interaction

As stated in Sect. 4.3, the global planner is able to interact with the local planner to request new routes in cases where the current route can not be continued due to road blockage due to road works, accidents, etc. To implement this interaction, specific LCM messages have been defined to request and send routes as well as to communicate different types of warnings to the HMI module.

The new defined messages are the following:

- `OSM_ROUTE_REQUEST`: This message includes the current vehicle pose (position and heading) and the destination.
- `OSM_ROUTE_RESULT`: The resulting nodes and the unique identifier of the computed route are included in this message.
- `OSM_CORRIDOR`: This message is used to communicate the computed road corridor based on the nodes that conform the route to the destination. It includes a ordered list of the control points of the right and left boundaries of the road corridor, as well as the centreline. Being the road corridor based on Bézier curves

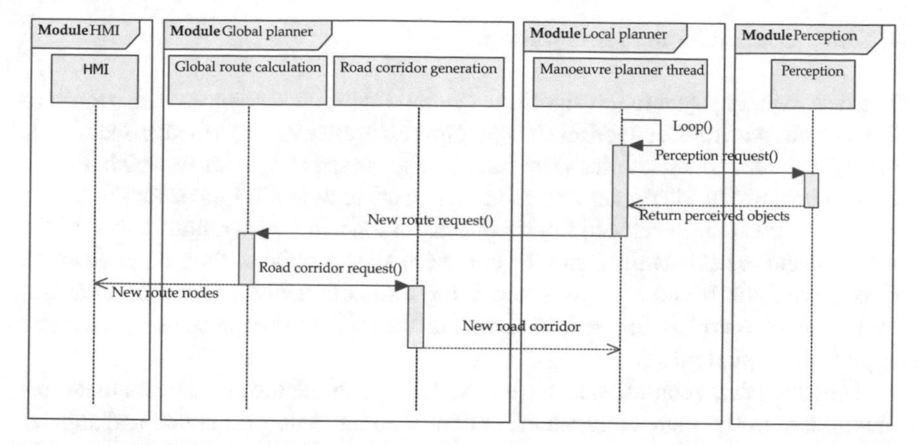

Fig. 7.9 UML sequence diagram of global and local planners and HMI interaction

the amount of information to communicate the road boundaries is considerably reduced in comparison with sending a full list of 2D points that define the raw road corridor.

- `WARNINGS`: When a warning needs to be issued to the HMI, this message is used to include information about the type the warning. Depending of the type of the warning different additional information can be also included. For example, when a dynamic obstacle is detected, the position, speed and motion direction can be provided.

As can be seen in the diagram of Fig. 7.9, besides the request of a new route to the global planner, a warning message is issued to the HMI module to alert the vehicle occupants of the road state and the future route change.

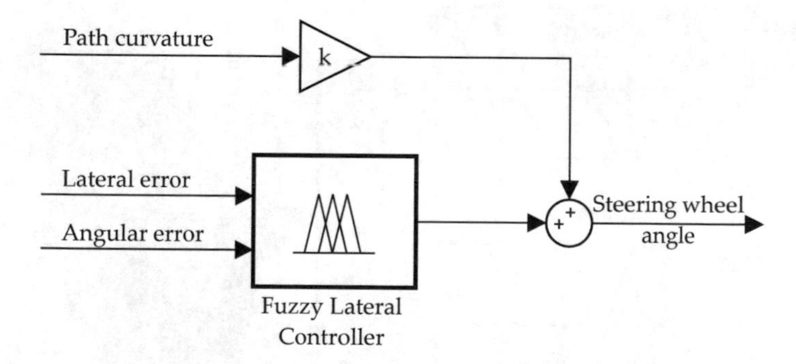

Fig. 7.10 Lateral control diagram

7.2.3.3 Control Module Adaptation

The new trajectory generation algorithms include substantial changes with respect to the previous one in both lateral and longitudinal controller. As described in Sect. 3.2.1, the prior generated trajectories were based on separated waypoints in which a value of the curvature in each point was estimated to serve as control input for the lateral controller. With the developed improvements in trajectory generation in this thesis, finer trajectories are obtained and the curvature at each point of the path is obtained through analytic Bézier equations, providing a smooth feed-forward control action to the lateral controller. Figure 7.10 shows the lateral control scheme used in the real experimental platform.

The lateral fuzzy controller had to be adapted to fit the new expected error measurements since the new smoother paths are translated into smoother lateral and angular error measurements. On the one hand, the new state estimation module introduces a high reduction of measured orientation noise so that the lateral controller could be adapted to expect lower and smoother values of the angular error. On the other hand, the same occurs to the lateral error. Being the followed paths smoother, the lateral error can be calculated in a more accurately manner. Based on this improvements in the input control signals, the fuzzy controller was adjusted obtaining the membership functions and control surface, shown in Fig. 7.11. In consequence, a smoother and more accurate path following performance have been achieved.

The longitudinal control scheme remains the same as in prior architecture (see Fig. 7.12). Nevertheless, the trajectories generated with the new local planner provide smoother speed reference values for each point of the planner trajectory. As a result, the fuzzy longitudinal controller also had to be adjusted since it was designed to work with step-like speed reference signals while the developed speed planning algorithm developed in this thesis is able to generated continuous speed reference values with

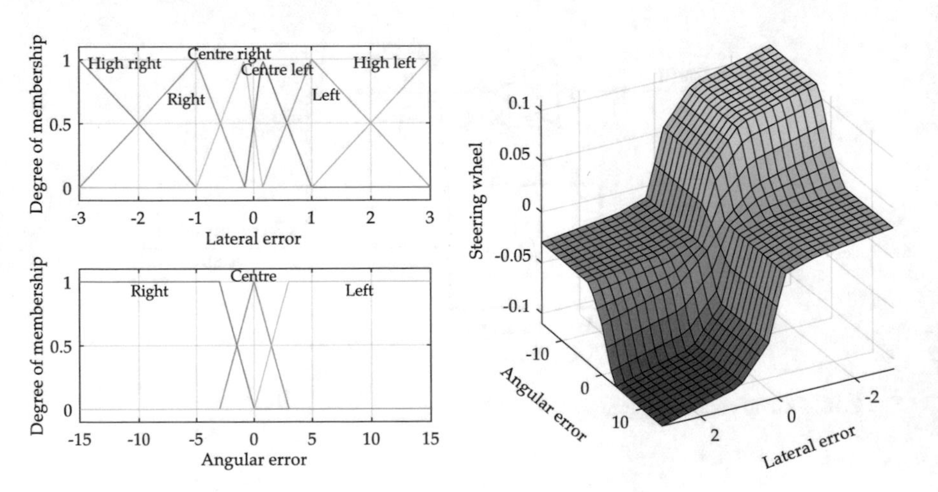

Fig. 7.11 Lateral fuzzy controller

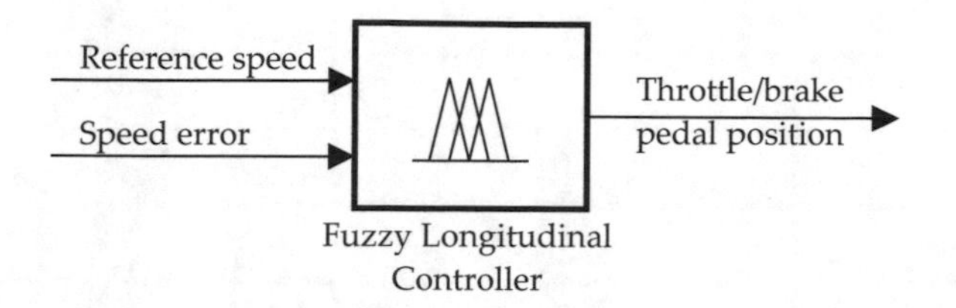

Fig. 7.12 Longitudinal control diagram

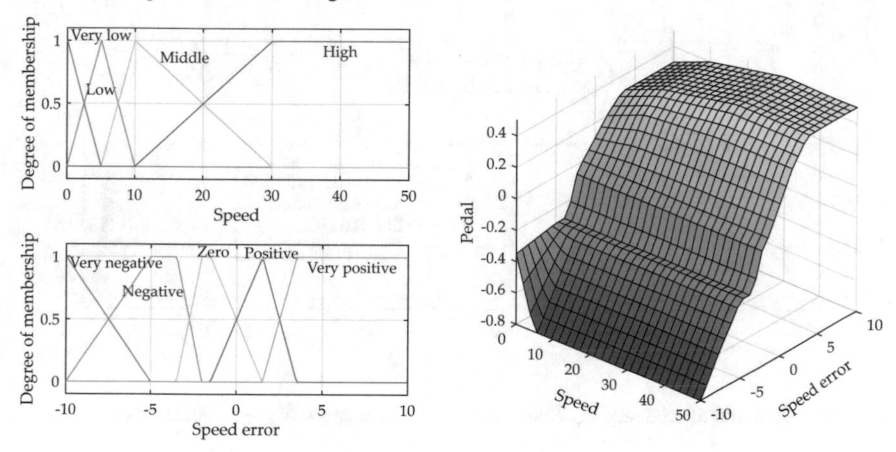

Fig. 7.13 Longitudinal fuzzy controller

limited longitudinal and lateral accelerations. The final membership functions and control surface of the longitudinal controller are shown in Fig. 7.13.

7.2.3.4 State Estimation Module

The goal of the state estimation module is to provide a reliable information about the state of the vehicle rather than relying only on raw GPS localization data. Different techniques for state estimation co-exist in the literature. One of the most used in mobile robotics is the Kalman Filter and variations [5, 11].

Bearing the above in mind, a Kalman Filter has been implemented to fuse information coming from different on-board sensors. Although GPS data remains the basis of the vehicle localization, CAN-bus data (vehicle speed, longitudinal acceleration and yaw rate) is used to improve the state estimation.

The estimation of the vehicle heading is the most influential variable on the lateral controller performance since its noise is transmitted directly to the angular error (being the angular error the difference between the heading and the path orientation) and proportionally to the lateral error (as the lateral error is computed with respect to the bumper position). One of the main improvements achieved with this module is the heading estimation even at low speeds. Figure 7.14 shows the raw speed and heading together with filtered ones during a test.

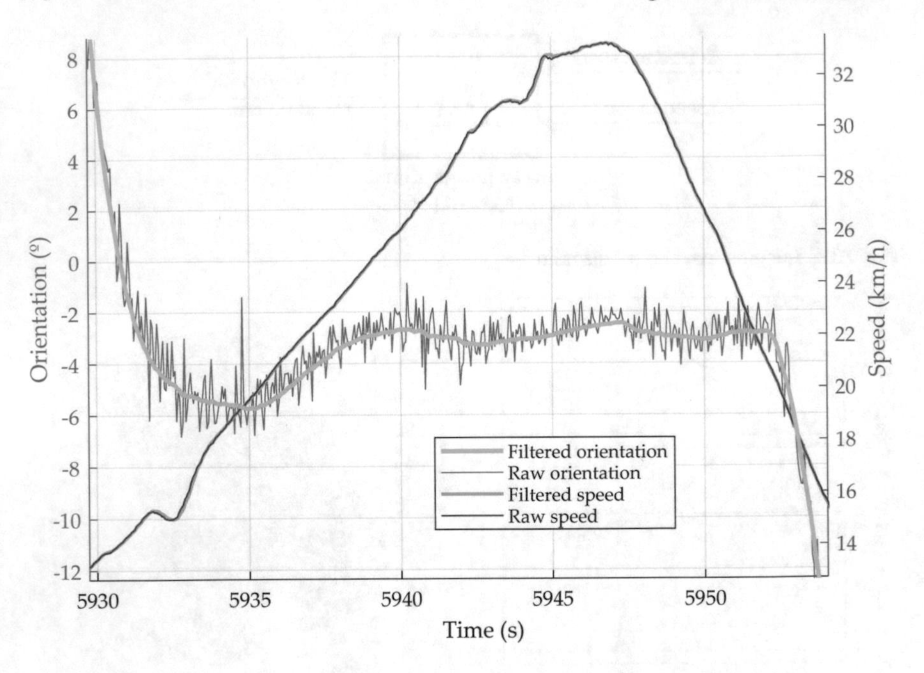

Fig. 7.14 Raw and filtered heading (left ordinate axis) and speed (right ordinate axis)

Summarizing, this module provides two main advantages:

- This module provides reliable information on the state of the vehicle. In this way, the modules that need this information do not need to apply any processing or filtering of the vehicle status.
- A better state estimation of the vehicle allows to better control it.

7.2.3.5 LiDAR-Based Object Detection

In order to identify objects in the vehicle surroundings through the LiDAR data, two different approaches have been developed: (i) a library of artificial intelligence-based methods for obstacle detection from raw LiDAR data, and (ii) a filtering algorithm on the objects data provided by the LiDAR sensor as described in Sect. 7.2. Since the implementation of the artificial intelligence-based library is still not feasible for a real-time setting, the objects data provided by the Ibeo processing unit have been used to perform the trials presented in this thesis.

Artificial intelligence-based methods for obstacle detection

Nowadays, many classifiers rely on machine-learning approaches to exploit data redundancy and abundance to find out patterns, trends and relations amongst input attributes and class labels [20]. Within obstacle-recognition techniques, vector support machines have been widely applied for classification and regression problems [22]. An interesting application using machine learning for pedestrian detection in autonomous vehicles based on High Definition (HD) 3D LiDAR is reported in [16], providing more accurate data to be successfully used in any kind of lighting conditions.

The library proposed is composed by three classification methods: a multi-layer perceptron neural network (MLP), a self-organizing map (SOM) and a support vector machine (SVM). The library is integrated into a co-simulation framework for obstacle recognition on the basis of sensory data provided by a virtual sensor network. This co-simulation framework is designed and built using SCANeR studio (a software suite dedicated to automotive simulation) and Matlab/Simulink as depicted in Fig. 7.15. Moreover, an assistance-driving scenario is created in SCANeR in order to emulate the real environment.

The whole system is evaluated in a particular use case built from two types of sensory data (LiDAR and GPS sensors) within the defined scenario. The comparative study demonstrates that the proposed obstacle detection methods are powerful strategies for pedestrian detection under good weather conditions.

In the validation study, six performance indices were considered: correct classify samples or correct rate (CCR), incorrectly classified samples or error rate (ECR), the mean absolute error (MAE), the root mean squared error (RMSE), the relative absolute error (RAE) and the root relative squared error (RRSE). In the training and validation phase of the classification models, the best results were achieved with the multi-layer perceptron and the support vector machine, but the self-organizing map did not perform so badly as to be discarded from future analyses. The results of the comparative study of the classifiers are summarized in Table 7.3.

The full details about this work can be found in [4], where further evaluation of the proposed library is also performed in the presence of different weather conditions (rain, fog and snow).

Although the results of the proposed object recognition and classification library obtained in a simulated environment under Matlab/Simulink are promising, the library is still not implemented to be run on-line in the experimental vehicle. Therefore, the objects data provided by the Ibeo processing unit have been used by the planning modules. In this connection, a software module has been developed to acquire and format the object data and sent them through a dedicated LCM channel.

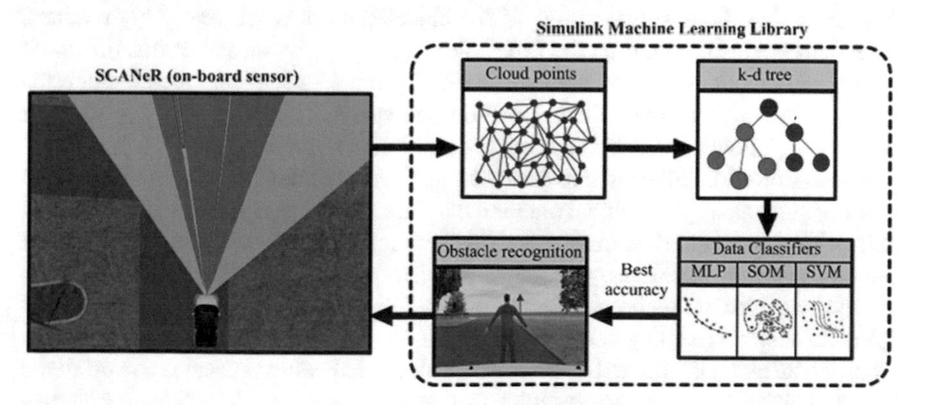

Fig. 7.15 Co-simulation framework set up for obstacle detection

Table 7.3 Comparative study of MLP, SVM and SOM classifiers under good weather conditions

Performance index	MLP	SVM	SOM
CCR (%)	88.19	91.36	90.91
ECR (%)	11.81	8.64	9.09
MAE	0.12	0.09	0.09
RMSE	0.34	0.29	0.30
RAE (%)	23.64	17.29	18.68
RRSE (%)	9.274	7.93	8.36

Rule-based object filtering

Although the object information provided by the LiDAR sensor processing unit contains a good starting point for estimating the state variables of objects in the environment around the vehicle, many of the initially detections are not significant or do not correspond to reality. Indeed, as this sensor has only 4 layers, the accuracy of the perceived objects' state is not always reliable enough. However, some cases of inaccurate objects estimation can be easily filtered to obtain a final set of relevant objects.

Since the goal of the LiDAR-based object detection is to provide reliable object information from the objects data to the motion planning algorithm of the architecture, different filtering rules have been defined to eliminate objects that are clearly outside of the road corridor or do not present a hazard for the vehicle (such as small bushes, buildings, walls, etc.).

The stated thresholds and their descriptions are presented in Table 7.4, where $ot_{min-dist}$ and $ot_{max-dist}$ allow to specify a region of interest. Concretely, $ot_{min-dist}$ is used to avoid suddenly detected objects very close to the car, assuming that the object would have been detected earlier if it were really there. The $ot_{min-edge}$ threshold makes possible to filter really small objects that do not present hazard for the vehicle. Moreover, the object data structure include the age of each object in seconds. In this way, $ot_{min-age}$ is used to ignore objects that have not been detected for a minimum amount of time, taking thus into account false positive detections in a few frames. Assuming that the absolute speed of the object can not reach really high values, detections are also limited by the threshold $ot_{max-speed}$. By means of this threshold many bad detections can be eliminated. Finally, the data structure also include the standard deviation of the vehicle position and speed. Therefore, $ot_{max-\sigma-pos}$ and $ot_{max-\sigma-pos}$ allow to identify bad object detections to exclude them.

In addition of the filtering threshold of Table 7.4, and taking advantage of the road corridor generated by the global planner, the static obstacles that are outside the road corridor are considered as irrelevant. This strategy allows to exclude obstacles that can be potentially detected by the sensor, such as big objects that are clearly outside the road corridor but its bounding box falls within it. This strategy is not applied to dynamic obstacles as they could be moving to the road corridor area e.g. pedestrian crossing the street or unidentified moving objects that could present a hazard to the vehicle safety. However, the obstacles that are outside the road corridor but exceed

Table 7.4 Threshold used to filter object data from LiDAR

Threshold	Description
$ot_{min-dist}$ and $ot_{max-dist}$	Minimum and maximum distance to the closest point of the obstacles (m)
$ot_{min-edge}$	Minimum length of object box edges (m)
$ot_{min-age}$	Minimum object age (s)
$ot_{max-\sigma-pos}$	Maximum value of standard deviation of object position (m)
$ot_{max-speed}$	Maximum object speed (m/s)
$ot_{max-\sigma-speed}$	Maximum value of standard deviation of object speed (m/s)
ot_{dyn}	Minimum speed threshold to consider an object as dynamic (m/s)

a defined speed threshold (ot_{dyn}) are also considered as dynamic objects. Thus, the algorithm is able to return a list of the relevant static and dynamic objects.

With the above in mind, let define and object structure (o) composed of objects information such that $o_i = (d_i, a_i, s_i, sx_i, sy_i, \sigma p_i, \sigma s_i)$, where d_i is the distance to object, a_i is the age of the object, s_i is the speed of the object, sx_i and sy_i are dimensions of the bounding rectangle, and σp_i and σs_i are the standard deviation of object position and speed, respectively. Then, the filter is applied in a general case with n objects as represented in Algorithm 1.

Input: Raw objects structure (o), Thesholds in Table 7.4
Output: Filtered objects
$o_{FS} \leftarrow$ Define empty structure for filtered static objects;
$o_{FD} \leftarrow$ Define empty structure for filtered dynamic objects;
foreach $i \leftarrow 1$ ***to*** n **do**
  if $d_i < ot_{max-dist} \land d_i > ot_{min-dist}$ **then**
    if $a_i > ot_{min-age}$ **then**
      if $s_i < ot_{max-speed}$ **then**
        if $sx_i < ot_{min-edge} \land sy_i < ot_{min-edge}$ **then**
          if $(\sigma p_i < ot_{max-\sigma-pos} \land \sigma s_i < ot_{max-\sigma-speed} \land s_i > ot_{dyn}) \lor$ *IsInsideRoadCorridor(o_i)* **then**
            if $s_i > ot_{dyn}$ **then**
              $o_{FD} \leftarrow o_i$;
            else
              $o_{FS} \leftarrow o_i$;
            end
          end
        end
      end
    end
  end
end
return o_{FS}, o_{FD};

Algorithm 1: Rule-based object filtering algorithm

where `IsInsideRoadCorridor(`o_i`)` is the function that checks if the object o_i is inside the road corridor, o_{FS} and o_{FD} are the data structures containing the static and dynamic objects, respectively.

The most computationally expensive section of the algorithm is the function `IsInsideRoadCorridor(`o_i`)`. Note that it is only called if the rest of thresholds have been verified, thus avoiding to check if the object is inside the corridor when the object is not considered as relevant.

7.2.3.6 Camera Acquisition Module

A software module for camera acquisition has been implemented to capture, process and send the image through LCM API as depicted in Fig. 7.7. Although this figure only shows one *image processing* module, the proposed architecture enables to include modules to extract different image features that can be sent through LCM. So far, the features include the lane detection information that is used to adapt the self-generated road corridors from OSM.

Given the large amount of information provided by the camera, the implementation of this module also foresees the sending of images through shared memory to modules that requires the raw image. This strategy will avoid the available bandwidth in the vehicle local network to be fully occupied by raw data sent through LCM.

7.2.3.7 HMI and Visualization Interfaces

In order to facilitate the vehicle-user interaction, three different tools have been developed. One of them provides a way for the user to select the destination point thought a friendly Android environment, while the others are dedicated exclusively to visualization (both a better situation awareness and for debugging purposes): (i) a 2D tool integrated with the control program and a (ii) 3D standalone tool that is fed from LCM information. Both are described in dedicated paragraphs below.

The implementation of visualization tools was motivated by the need of interpreting and understanding the vehicle state, the perceived environment and the outputs of the software components of the architecture. These tools have made easier the development and debugging tasks needed in the implementation stage of the algorithms presented in this thesis.

Android tablet interface

An on-board Android human-machine interface relying on the open source project *OsmAnd* [18] has been implemented. This application is able to work with OSM data and provide some general information about the global route the vehicle is following: the route itself, estimated time to destination, the vehicle speed, etc.

To adapt the original Android application to the decision-making architecture needs, some modification have been carried out over the existing application:

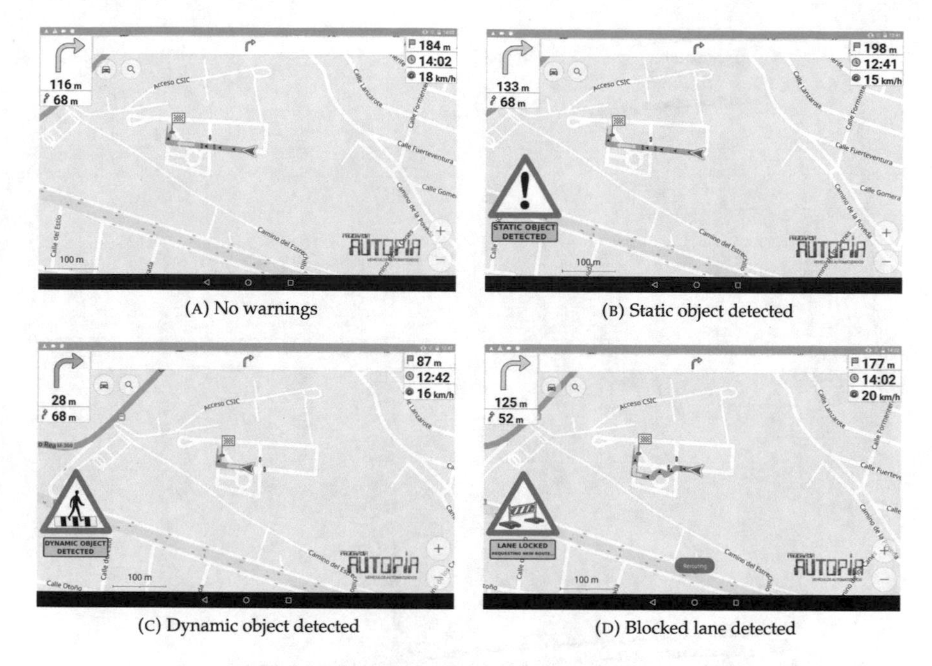

(A) No warnings

(B) Static object detected

(C) Dynamic object detected

(D) Blocked lane detected

Fig. 7.16 Screen shots of the Android application issuing warnings in specific situations

- An Android service has been developed to enable the communication through LCM from the Android tablet using wired connection. Thus, the application is able to send and receive information from other architecture modules.
- A new option to select the global planner used in the new architecture (see Sect. 4.3) has been added. This option allows to activate the sending of route requests to the global planner module, that is run in the on-board PC.
- The application has been modified to show the most relevant information perceived by the embedded sensors. Thus, when a static/dynamic obstacle or a blocked lane is detected, a specific warning is shown in the screen (see Fig. 7.16).

2D visualization interface

This interface was initially developed to represent the planned trajectories as well as the vehicle historic positions. This made it easier to evaluate the performance of the controllers while the vehicle is following a trajectory. However, more information has been subsequently added. It was designed to be integrated with the control-planning application of the new architecture to avoid sending data for visualization through IPC, using valuable network bandwidth.

Different kind of data are included in the interface. Firstly, the raw GPS data and the localization data provided by the state estimation module are used to locate the vehicle with respect to a global coordinate system. In addition, the road corridors generated by the global planning components are also plotted. Regarding the perceived static objects, only the objects of interest selected after the application of the rule-based

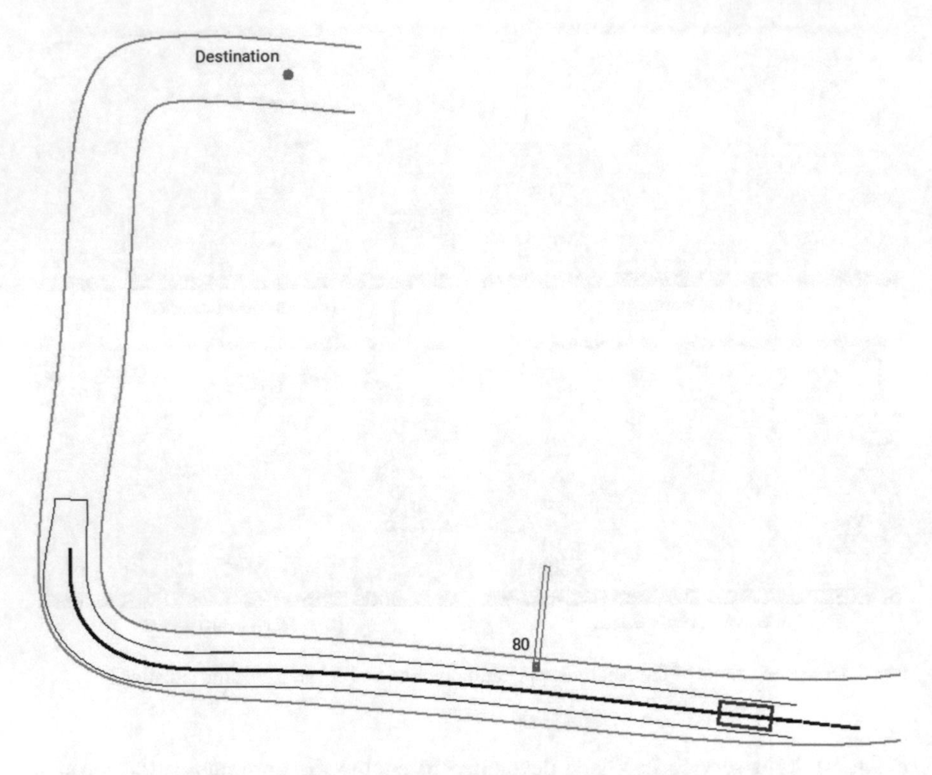

Fig. 7.17 2D visualization interface screen shot while a dynamic obstacle is detected

filter described in Sect. 7.2.3.5 are represented. With respect to dynamic objects, their current location as well as the predicted path are shown in this interface as can be seen in Fig. 7.17. Thus, the objects taken into account by the planning algorithms are shown in runtime.

The output of local planner is also plotted in the 2D visualization: the current path that the vehicle is following is plotted. Furthermore, the path-polygon is also delineated for debugging purposes.

3D visualization interface

A 3D visualization tool is useful when the amount of data to be interpreted by the user becomes high. A firstly developed simulation and visualization tool [1–3] based on Matlab/Simulink was useful to simulate and reproduce trials carried out with the real test platform. However, the need of representing more information such as occupancy grids or perceived objects motivated the development of a new tool dedicated exclusively to visualization.

The application is based on RViz, a visualization package included in ROS libraries [21]. Since the communication among the different software components of the AUTOPIA architecture is based on LCM and RViz needs ROS messages, the

Fig. 7.18 3D visualization interface screen shot

implemented tool receives LCM messages containing information to be represented and generates similar ROS messages that are sent to Rviz.

Besides the vehicle itself (a 3D model of the vehicle is included), useful data is also visualized. On the one hand, perceived information is represented: the localization of the vehicle is used to locate the vehicle with respect to a global coordinate system, where the road corridors are also represented. On the other hand, both intermediate and final computations of architecture algorithms are represented. Firstly, the probabilistic occupancy grid used to place the localization uncertainty dependent road corridor and the perceived objects is showed. Moreover, the final computed path is also visualized. This tool is useful to understand the information coming from the sensors (localization, perceived objects).

The represented information in this tool is: the vehicle localization from the state estimation module, the perceived objects by the LiDAR sensor, the road corridors generated by the global planner components, the probabilistic occupancy grid and the current planned path. All this information is updated in runtime. A screenshots is shown in Fig. 7.18.

This interface offers some advantages over the 2D visualization tool described above. One of them is that it can be run in a different computer connected to the vehicle network. As a result, the resources of the on-board PC dedicated to control the vehicle are not employed to create computationally expensive visualizations. Moreover, the use of a 3D visualization improves the situation awareness for the user by enabling to set specific points of view in a three dimensional world.

7.2.3.8 Architecture Profiling

Since some of the modules in the architecture require considerable computational resources, it is important to quantify their computing times in order to know the approximate response times and to detect possible bottlenecks. To this end, different measurements of execution times have been carried out for each of the algorithms of the architecture. The results of the approximated computing times ranges are summarized in the Table 7.5.

Note that the computing time of the global route calculation and the road corridor generation depend on route length. For large routes these times increase. The computing time of the localization uncertainty propagation algorithm depends on the number of cells of the grid, as shown in Sect. 6.5.2.

In the case of the risk estimation algorithm, the calculation time depends to a large extent on the following factors: (i) the number of traffic agents that are around the ego-vehicle, and (ii) the number of particles used in the performance of the algorithm. The computing time range shown in Table 7.5 for this algorithm, has been obtained by using 1600 particles in the test cases presented in Sect. 5.3.3.

As can be seen in Table 7.5, the most computationally expensive algorithms are the risk estimation algorithm, the localization uncertainty propagation over the grid, and the trajectory generation when using occupancy grid (as described in Sect. 6.5). The timing performance of these algorithms can be improved by using parallel computing in their implementations.

Table 7.5 Results of computing time measurements of the implemented algorithms

Algorithm	Module	Computing time (ms)
Global route calculation algorithm	Global route calculation	1–3
Road corridor generation	Road corridor generation and adaptation	5–20
Vision-based road corridor adaptation	Road corridor generation and adaptation	25–35
Occupancy grid: localization uncertainty propagation	Road corridor generation and adaptation	5–100
Risk estimation algorithm	Motion prediction	100–200
Simplified motion prediction	Motion prediction	<1
Manoeuvre planner algorithm	Manoeuvre planner	<1
Trajectory generation algorithm	Trajectory generation	15–75
Trajectory generation (using occupancy grid)	Trajectory generation	110–200

7.3 Demonstrations

In addition to specific validation performed for each of the architecture modules presented in Chaps. 4–6, the proper operation and integration of the whole architecture has been verified through several live demonstrations in two different events.(i) *2018 IEEE/RSJ International Conference on Intelligent Robots and Systems* (IROS 2018), that was held in Madrid from 1st to 5th October 2018, and (ii) *S-Moving 2018* event, which was held in *Palacio de Ferias y Congresos de Málaga* (FYCMA), Malaga.

Both demonstrators showed the capabilities of the architecture proposed in this thesis through different scenarios, using the fully automated Citroën DS3 described in Sect. 7.2.

7.3.1 Demonstration at IROS 2018

This demonstration took place in the *Autonomous Driving Events* section of the *2018 IEEE/RSJ International Conference on Intelligent Robots and Systems* (IROS 2018) that was held at the Centre for Automation and Robotic (CSIC-UPM) facilities in Arganda del Rey, Madrid. The test track used in the demonstration mimics an urban environment including tight curves, straight sections, several intersections, a roundabout and traffic lights (see Fig. 7.19).

Within this event, five participant teams coming from different institutions—Consejo Superior de Investigaciones Científicas (CSIC—AUTOPIA), Universidad Carlos III de Madrid (UC3M), Universidad Alcalá de Henares (UAH—*Robesafe* and *INVETT*), Institut National de Rechcrchc cn Informatique et Automatique (INRIA)—performed several exhibitions focusing on one or several of the following aspects:

Fig. 7.19 Test track at the Centre for automation and robotics (CSIC-UPM), Arganda del Rey, Madrid

- Low speed (20 km/h) or high speed (40–50 km/h)
- Perception (static and moving obstacles, road signs, etc.)
- Localization & mapping
- Control, planning & decision-making
- V2X-based cooperative systems (merging, intersections, roundabouts management)
- V2I cooperation with traffic lights system
- Driver-vehicle interaction.

The AUTOPIA demonstration was entitle "OSM-based navigation and decision-making". It showcased the proposed navigation architecture in this thesis by using the experimental platform described in Sect. 7.2. To that end, different trials ranging from dynamic global planning to local motion planning, including obstacle avoiding, were performed. The static and dynamic obstacles were detected on the road by a multi-layer LiDAR, the planner evaluated whether an avoiding manoeuvre was feasible in the pre-computed route, or an automatic re-routing request could be performed.

The live demonstration of AUTOPIA consisted of five different scenarios that used the proposed decision-making architecture in this thesis:

- **Scenario 1**: Low speed driving
- **Scenario 2**: High speed driving
- **Scenario 3**: Static obstacle avoidance
- **Scenario 4**: Dynamic object avoidance
- **Scenario 5**: Automatic re-routing due to road blockage.

In scenarios 1 and 2, predefined trajectories were used to show the capabilities of the vehicle control while overcoming tight curves and driving at high speeds, respectively. In contrast, scenarios 3, 4, and 5, the motion planning capabilities of the architecture: global planning from the initial vehicle pose to the given destination (performing re-routing when needed), and local planning to overcome the obstacles encountered while the vehicle is on its way to its destination.

In addition to the on-board decision system, V2I communications were used to show the cooperation with traffic lights system installed in the four-lanes intersection placed in the central point of the test track. In the five trials carried out, an intersection management system connected to the traffic lights was aware of the position and direction of travel of the vehicle and prioritized the vehicle passing by changing the lights states when the vehicle was approaching to the intersection from any of its ways.

Figure 7.20 shows a external picture and screen shot of the HMI and frontal vehicle image while driving automatically at 50 km/h during the second scenario.

The scenario 3 consisted on travelling from one part of the track to another, through a route where two static obstacles were hindering the vehicle driving. Figure 7.21 shows a picture taken when the vehicle was avoiding two consecutive static obstacles placed in the road.

Picture on Fig. 7.22 belongs to scenario 4. It shows the frontal image of the vehicle when it was adapting the speed profile of its trajectory to avoid the collision with a pedestrian that is crossing the street. The collision avoidance was performed successfully.

Finally, the last scenario comprised a road blockage in the route that the vehicle was following to reach the destination as shown in Fig. 7.23A. When the perception system of the vehicle detected the static obstacle that was blocking the road, the local planner requested a new high-level route to the global planner (see Fig. 7.23B), that was quickly retrieved, causing a smooth transition that was not even noticed inside the vehicle.

The full demonstration was retransmitted in streaming achieving a high diffusion of the event. Moreover, a complete video of the whole demonstrator can be accessed

(A) HMI and frontal vehicle image

(B) Picture of the vehicle while driving at 50 km/h

Fig. 7.20 High speed automated driving demonstration while performing the second scenario

Fig. 7.21 Scenario 3: two consecutive static obstacles being avoided

Fig. 7.22 Frontal vehicle image in the scenario 4 while avoiding a pedestrian that is crossing the street

(A) Obstacles blocking the road at the central intersecion in scenario 5.

(B) Re-routing due to road blockage. Blocking obstacle detected and re-routing request

(C) Re-routing due to road blockage. Route retrieved

Fig. 7.23 Frontal vehicle image and HMI screenshots when a static object that is blocking the lane is detected

in the following link: https://youtu.be/SGVnmsUS06k. Furthermore specific links to each of the demonstrated scenarios are provided below:

- **Scenario 1**: Low speed driving. Link: https://youtu.be/SGVnmsUS06k?t=225
- **Scenario 2**: High speed driving. Link: https://youtu.be/SGVnmsUS06k?t=358
- **Scenario 3**: Static obstacle avoidance. Link: https://youtu.be/SGVnmsUS06k?t=527
- **Scenario 4**: Dynamic object avoidance. Link: https://youtu.be/SGVnmsUS06k?t=667
- **Scenario 5**: Automatic re-routing due to road blockage. Link: https://youtu.be/SGVnmsUS06k?t=811.

7.3.2 Demonstration at S-Moving 2018

The *S-Moving* event is a space of reference for companies, professionals, entities and public administrations related to intelligent, autonomous and unmanned vehicles sectors. It was held in the *Palacio de Ferias y Congresos de Málaga* (FYCMA), Málaga (Spain), from 17th to 18th October 2018. This event is an international forum dedicated to technologies applied to intelligent, autonomous, connected and unmanned mobility by land, sea and aerospace and their infrastructures.

AUTOPIA Program had the opportunity to make live demonstrations during the event. Just like in the IROS 2018 demonstration (see Sect. 7.3.1), the architecture developed in this thesis was used to exhibit its automated driving capabilities in different situations (Fig. 7.24).

Within the driving area (see Fig. 7.25), the demonstration scenario consisted on a trip from one extreme to the other while some static and dynamic obstacles were hindering the way until the destination point.

With respect to the IROS 2018 demonstration, the scenario exhibited in the *S-Moving* event presented two additional difficulties. On the one hand, the constrained driving area caused a greater obstacle density during the track so that three obstacles were consecutively placed: two static and one dynamic. On the other hand, the

(A) AUTOPIA vehicle in the driving area of S-moving event

(B) AUTOPIA vehicle and static obstacle used in the demonstation

Fig. 7.24 AUTOPIA vehicle in the FYCMA facilities at the *S-moving 2018* event

dynamic obstacle was a pedestrian that stopped walking in the middle of the street while the vehicle was driving automatically. This included a higher scenario complexity as the dynamic obstacle changed to be static and the decision-making system had to interpret the situation to find a different way to overcome the same obstacle. A scheme of the scenario is shown in Fig. 7.26.

The most critical point of this demonstration was the pedestrian avoidance since it firstly was moving at constant speed and then stopped abruptly in the middle of the street. To show the output of the decision-making system at this point, three screenshots of the 3D visualization of the demonstration are presented in Fig. 7.27. In these screenshots the pedestrian is shown as a small blue rectangular parallelepiped close to the vehicle path, that is drawn in magenta.

The path planned when the pedestrian was walking and consequently detected as a dynamic object, is shown in Fig. 7.27A. In contrast, Fig. 7.27B shows the planned

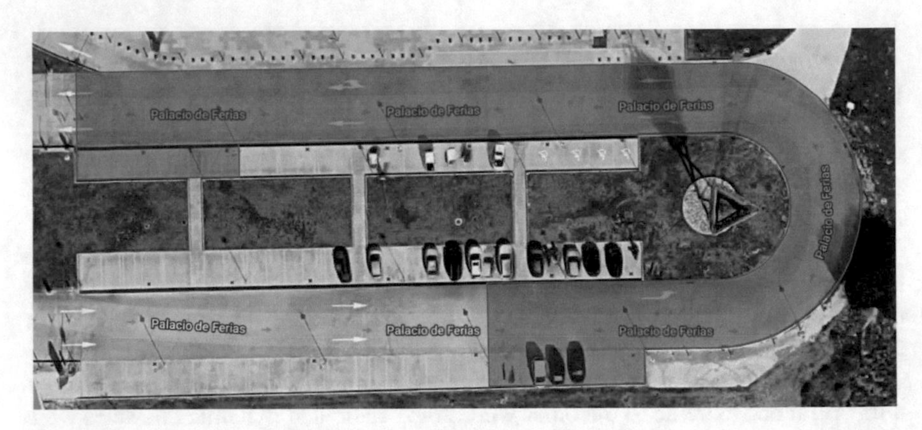

Fig. 7.25 The space reserved for the live demonstration is shaded red area

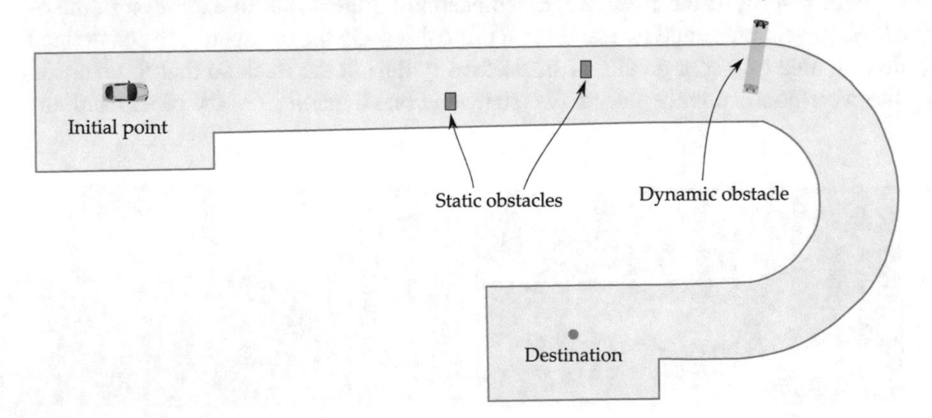

Fig. 7.26 Demonstration scenario at *S-moving 2018* event

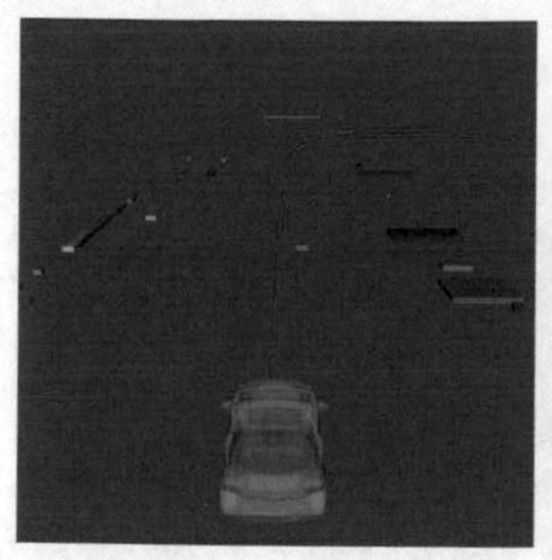

(A) The pedestrian crossing the street was detected and the vehicle adapted its speed to avoid de collision

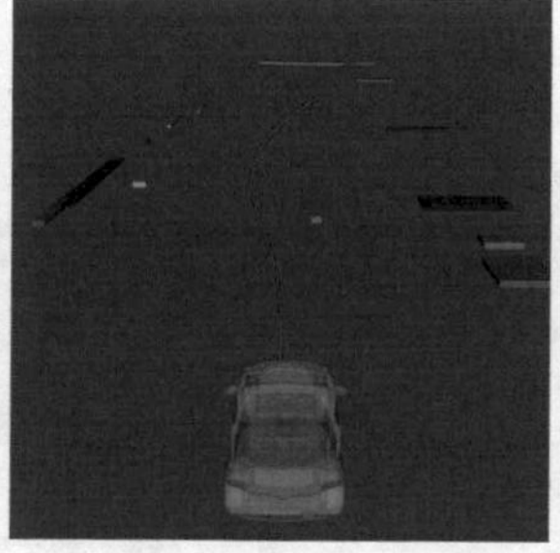

(B) The obstacle suddenly stopped in the middle of the street and instantly the vehicle planned a different trajectory to avoid it.

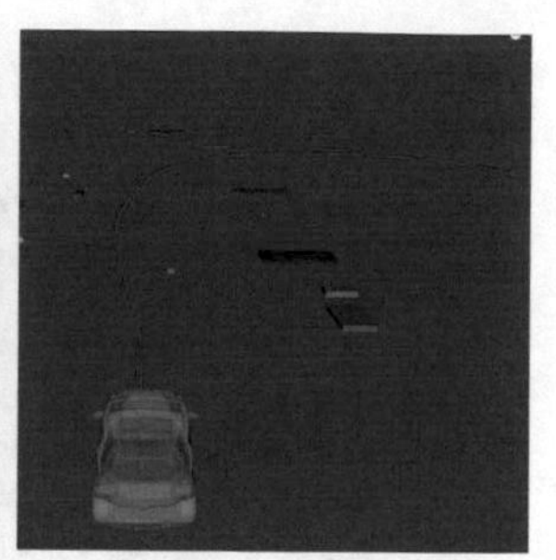

(C) The new trajectory is extended.

Fig. 7.27 3D visualization of the planned trajectory and perceived obstacles when the dynamic obstacle is being avoided

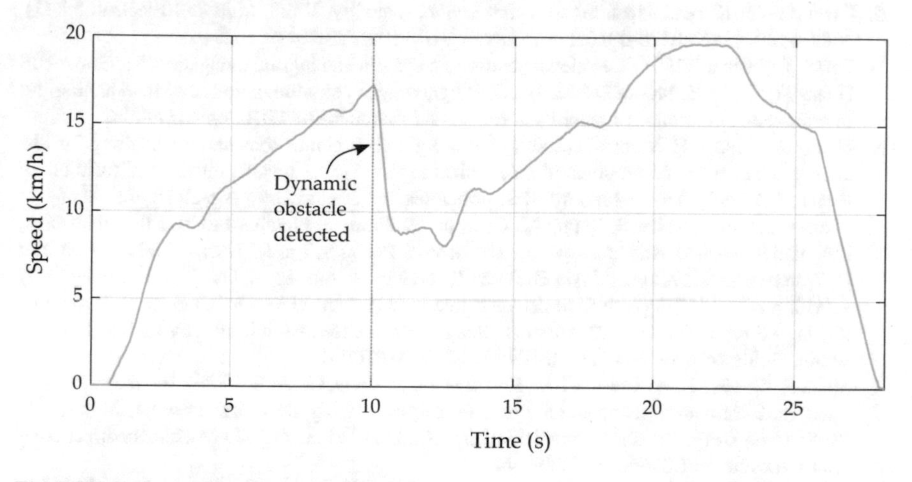

Fig. 7.28 Vehicle speed during the demonstration

path when the pedestrian stopped and was detected as a static obstacle. Finally, Fig. 7.27C shows how the new trajectory that avoids the obstacle is extended.

In Fig. 7.28, the vehicle speed during the demonstration is shown. As can be noticed, at instant $t = 10$ s, the dynamic obstacle is detected. Consequently, the speed of the vehicle is reduced to avoid the collision.

This driving scenario was performed several times during the event, and all of the demonstrations were successfully completed.

This event was covered by national and regional TV media (RTVE and Canal Sur). A live recording of one of the demonstration performance can be found in the following link: https://youtu.be/Q9Oskis7iR0.

References

1. Artuñedo A, Godoy J, Haber R (2016) Entorno avanzado de co-simulación para maniobras cooperativasentre vehículos. In: Actas de las XXXVII Jornadas deAutomática. Comité Español de Automática(CEA-IFAC), Madrid, Sept 2016, pp 704–709. ISBN: 978-84-617-4298-1. http://ja2016.uned.es/assets/files/ActasJA2016.pdf
2. Artunedo A, Godoy J, Haber R, Villagra J, del Toro RM (2015) Advanced co-simulation framework for cooperative maneuvers among vehicles. In: 2015 IEEE 18th international conference on intelligent transportation systems. IEEE, Sept 2015, pp 1436–1441. ISBN: 978-1-4673-6596-3. https://doi.org/10.1109/ITSC.2015.235. https://ieeexplore.ieee.org/document/7313327/
3. Artuñedo A, Haber R, Matía F (2016) A co-simulation environment for cooperative maneuvers among vehicles. In: Industrials research meeting 2016. ETSII, UPM, Madrid, Apr 2016, p 219. ISBN: 978-84-16397-31-0. http://oa.upm.es/40073/
4. Castaño F, Beruvides G, Haber R, Artuñedo A (2017) Obstacle recognition based on machine learning for on-chip LiDAR sensors in a cyber-physical system. Sensors 17(9), 2109. ISSN: 1424-8220. https://doi.org/10.3390/s17092109. http://www.mdpi.com/1424-8220/17/9/2109
5. Chen SY (2012) Kalman filter for robot vision: a survey. IEEE Trans Ind Electron 59(11), 4409–4420. ISSN: 0278-0046. https://doi.org/10.1109/TIE.2011.2162714
6. Farrell J, Barth M (1999) The global positioning system and inertial navigation. McGraw- Hill
7. Huang A, Olson E, Moore D (2010) LCM: lightweight communications and marshalling. In: International conference on intelligent robots and systems, Oct 2010, pp 4057–4062
8. Huang AS, Olson E, Moore D (2009) Lightweight communications and marshalling for low latency interprocess communication. Technical report, Massachusetts Institute of Technology
9. Kaplan ED (1996) Understanding GPS: principles and applications. Artech House
10. Leonard J, How J, Teller S, Berger M, Campbell S, Fiore G, Fletcher L, Frazzoli E, Huang A, Karaman S, Koch O, Kuwata Y, Moore D, Olson E, Peters S, Teo J, Truax R, Walter M, Barrett D, Epstein A, Maheloni K, Moyer K, Jones T, Buckley R, Antone M, Galejs R, Krishnamurthy S, Williams J (2009) A perception-driven autonomous urban vehicle. In: Buehler M, Iagnemma K, Singh S (eds) The DARPA urban challenge: autonomous vehicles in city traffic. Springer, Berlin, Heidelberg, pp 163–230. ISBN: 978-3-642-03991-1
11. Matía F, Jiménez A, Al-Hadithi BM, Rodríguez-Losada D, Galán R (2006) The fuzzy Kalman filter: state estimation using possibilistic techniques. Fuzzy Sets Syst 157(16), 2145–2170. ISSN: 0165-0114. https://doi.org/10.1016/j.fss.2006.05.003. http://www.sciencedirect.com/science/article/pii/S0165011406002144
12. Milanes V, Godoy J, Perez J, Vinagre B, Gonzalez C, Onieva E, Alonso A (2010) V2I-based architecture for information exchange among vehicles. In: 7th symposium on intelligent autonomous vehicles, 43(16), 85–90. https://doi.org/10.3182/20100906-3-IT-2019.00017. http://www.sciencedirect.com/science/article/pii/S1474667016350376
13. Milanes V, Villagra J, Godoy J, Simo J, Perez J, Onieva E (2012) An intelligent V2I-based traffic management system. IEEE Trans Intell Transp Syst 13(1), 49–58. ISSN: 1524-9050. https://doi.org/10.1109/TITS.2011.2178839. http://ieeexplore.ieee.org/document/6121906/
14. Mitchell M, Oldham J, Samuel A (2001) Advanced Linux programming. Landmark series. New Riders. ISBN 9780735710436
15. Naranjo J, Gonzalez C, Garcia R, DePedro T, Haber R (2005) Power-steering control architecture for automatic driving. IEEE Trans Intell Transp Syst 6(4), 406–415. ISSN: 1524-9050. https://doi.org/10.1109/TITS.2005.858622. http://ieeexplore.ieee.org/document/1549844/
16. Navarro P, Fernández C, Borraz R, Alonso D (2016) A machine learning approach to pedestrian detection for autonomous vehicles using high-definition 3D range data. Sensors 17(12), 18. ISSN: 1424-8220. https://doi.org/10.3390/s17010018. http://www.mdpi.com/1424-8220/17/1/18
17. NMEA 0183—-Standard for interfacing Marine electronic devices. National Marine Electronics Association (U.S.) 2002

18. Offline mobile maps and navigation. https://osmand.net/
19. Perez J, Gonzalez C, Milanes V, Onieva E, Godoy J, de Pedro T (2009) Modularity, adaptability and evolution in the AUTOPIA architecture for control of autonomous vehicles. In: IEEE international conference on mechatronics (ICM), Apr, pp 1–5
20. Precup R-E, Angelov P, Costa BSJ, Sayed-Mouchaweh M (2015) An overview on fault diagnosis and nature-inspired optimal control of industrial process applications. Comput Ind 74, 75–94. ISSN: 01663615. https://doi.org/10.1016/j.compind.2015.03.001 https://linkinghub.elsevier.com/retrieve/pii/S0166361515000469
21. Quigley M, Gerkey B, Conley K, Faust J, Foote T, Leibs J, Berger E, Wheeler R, Ng A (2009) ROS: an open-source robot operating system. In: Proceedings of the IEEE international conference on robotics and automation (ICRA) workshop on open source robotics. Kobe, Japan, May 2009
22. Taghavifar H, Mardani A, Karim Maslak H (2015) A comparative study between artificial neural networks and support vector regression for modeling of the dissipated energy through tire-obstacle collision dynamics. Energy 89, 358–364. ISSN: 03605442. https://doi.org/10.1016/j.energy.2015.05.122. https://linkinghub.elsevier.com/retrieve/pii/S0360544215007446
23. Toulminet G, Boussuge J, Laurgeau C (2008) Comparative synthesis of the 3 main European projects dealing with cooperative systems (CVIS, SAFESPOT and COOPERS) and description of COOPERS demonstration site 4. In: 2008 11th international IEEE conference on intelligent transportation systems. IEEE, Oct 2008, pp 809–814. ISBN: 978- 1-4244-2111-4. https://doi.org/10.1109/ITSC.2008.4732652. http://ieeexplore.ieee.org/document/4732652/
24. Uzcategui R, Acosta-Marum G (2009) Wave: atutorial. IEEE Commun Mag 47(5):126–133
25. van Nunen E, Kwakkernaat MRJAE, Ploeg J, Netten BD (2012) Cooperative competition for future mobility. IEEE Trans Intell Transp Syst 13(3), 1018–1025. ISSN: 1524-9050. https://doi.org/10.1109/TITS.2012.2200475

Chapter 8
Conclusions

8.1 General Conclusions

The growing interest in ever higher levels of automation involves the development of algorithms capable of making decisions to face increasingly complex driving situations in a safe and human-like manner. In this regard, this thesis addresses the problem of motion planning in urban environments by proposing a decision-making and planning architecture where several modules with different functionalities provide human-like trajectories to the vehicle according to the perceived situation.

The proposed architecture improvements aims at pushing the navigation capabilities of automated vehicles when only non detailed and open-source maps are available. To that end, road corridors are dynamically obtained from these maps. To cope with their intrinsic uncertainty and low-fidelity, a camera-based lane detection system updates and enhances the navigable space, which, in addition, explicitly considers localization uncertainty. From that point, an efficient and human-like local planner determines, under a probabilistic framework, a safe motion primitive. LiDAR-based perception is then used to identify the driving scenario, and eventually re-plan the local path and speed profile, leading in some cases to re-adapt the route to be followed. These new functionalities represent significant progress beyond previous architecture of the AUTOPIA Program.

Extensive tests on real environments and different live demonstrations proved the robustness of the proposed architecture when dealing with different complex situations such as static and dynamic obstacle avoidance or dynamic re-routing when the high-level route is blocked.

A. Artuñedo, *Decision-making Strategies for Automated Driving in Urban Environments*, Springer Theses, https://doi.org/10.1007/978-3-030-45905-5_8

8.1.1 Contributions

This thesis proposes significant contributions to the prior existing AUTOPIA architecture for automated driving. The summarized specific contributions are listed below:

- **Global planning capabilities**: Different global planning functionalities have been added to the architecture. Firstly, a global routing based on OSM has been developed to provide a high-level definition of the route for the vehicle to reach its destination. Furthermore, an automatic road corridor generation algorithm has been developed to provide the available navigable space to the local planner. Moreover, a vision-based road corridor adaptation algorithm has been proposed to deal with possible OSM inaccuracy. In this context, two different approaches have been proposed to use the adapted road corridors in the motion planning strategy: (i) Directly use the generated and adapted road corridor assuming good enough localization, or (ii) a probabilistic occupancy grid approach that deals with localization uncertainty. This latter approach proposes a more general way to fuse data coming from maps and perception systems on the occupancy grid. Moreover, a localization uncertainty propagation algorithm over the grid has been proposed. Thus, the motion planning algorithms can be influenced by the current localization uncertainty of the vehicle.
- **Risk estimation and motion prediction**: Through the proposed risk estimation algorithm, which can be classified as an intention-aware prediction approach, it is possible to differentiate between the road agents close to the vehicle that may compromise the ego-vehicle safety. As a result, trajectory prediction algorithms can be applied only to agents that are causing possible risk situations, considerably reducing the number of predictions improving their accuracy, and therefore reducing the complexity of decision-making and planning. This is especially relevant in urban environments, where typically there is a large number of moving objects around the ego-vehicle, but many of them do not contribute to the generation of risk situations.
- **Local planning capabilities**: Regarding local planning, an extensive comparison of state-of-the-art path planning strategies for automated driving was firstly carried out. The results of the comparison allowed to choose the most appropriate planning strategy among the approaches assessed, taking into account a balance between calculation time and evaluated search space size. Moreover, the speed planning algorithm computes the appropriate speed for each point of the generated path considering limited longitudinal and lateral accelerations to ensure comfort inside the vehicle. On the other hand, the manoeuvre planning functionality gives the decision system the ability to select the appropriate planning mode according to the interpretation of the environment. The different proposed local planning modules have been successfully tested in a wide range of real scenarios where both static and dynamic obstacles had to be avoided.
- **Interaction capabilities**: To facilitate the interaction capabilities of the decision-making architecture and the vehicle occupants, a human-machine interface has

been implemented. This interface provides a friendly bidirectional communication channel that allows the occupants to set the destination point while the most relevant information about the trip (such as the route, time to destination, vehicle speed, etc.) is shown in a screen. In addition, the vehicle occupants can make use of the 2D and 3D visualization interfaces to show detailed information about the perceived environment and the decisions of the vehicle.

- **Architecture integration and validation**: All the algorithms developed for the different modules of the new architecture have been implemented and validated in a real experimental platform. For this purpose, different scenarios have been used in real operating environments and several live demonstrations have been carried out for the general public.

8.2 Future Work

The contributions of this thesis give rise to continuations of the work presented in the different research topics treated. The following list present future work lines grouped by topic:

- **Motion planning**: Despite the proposed trajectory generation algorithm offers good results in terms of trajectory quality and computation time, machine learning algorithms can be applied to enable a better use of the available computational resources when the computation time plays a critical role. This approach would lead to a prior selection of the path candidates to be evaluated, probably increasing the percentage of valid candidates by avoiding the evaluation of know-invalid candidates. Thus, besides saving computational resources, a better search space exploration or a more reduced planning time could be achieved. Regarding the speed planning, the comfort inside the vehicle could be improved by introducing methods for jerk limitation.
- **Probabilistic occupancy grid enrichment**: Currently, the occupancy grid includes the propagation of the localization uncertainty from the road corridors, and raw perceived obstacles. Future work around the occupancy grid representation will focus on objects identification and tracking from raw sensor data. This approach will allow to obtain a better perception of the near vehicle environment by fusing data coming from different perception sensors in the same grid. Furthermore, the uncertainty of the perceived objects location could be also addressed being propagating over the grid just like the localization uncertainty. These enrichments in perception would lead to a more efficient and human-like motion planning by avoiding unneeded consecutive planning requests caused by the bad perceived position of obstacles. Moreover, to improve the performance of the computation of the grid, parallel computing on GPUs should be used. In this connection, the right balance between computation time and grid cell size will be explored to increase the grid accuracy.

- **Risk assessment and decision-making**: The proposed risk estimation algorithm based on DBN can be extended for aggregated longitudinal and lateral collision risk estimation with other vehicles. Moreover, risk estimation with pedestrians, which so far are not considered because of the difficulty of finding decent prediction models, could be also included. This algorithm can be computationally expensive, so parallel computing on GPUs can be also used. Regarding the decision-making strategies, the use of Markov Decision Processes (MDP) could be explored to strengthen the current decision scheme.
- **Motion prediction**: The stability and reliability of the motion prediction of other traffic agents can be improved by applying multiple object tracking method using Kalman filter that takes as input reliable information of the perceived obstacles coming from clustering and tracking algorithms applied over the occupancy grid for object detection. Besides, the combination of probabilistic and machine learning techniques such as stochastic reachable sets and Gaussian mixtures could be applied to improve the motion predictions.

8.3 Dissemination

During the development of this thesis, the doctoral candidate has participated in different research projects, both national and European. This framework has allowed the establishment of working and collaboration links with other research centres, universities and related companies. This led to a 3-months research stay performed at the "Integrated Vehicle Safety" department at *Netherlands Organisation for Applied Scientific Research* (TNO), Nederlands, from to 17th April 2017 to 17th July 2017, under the supervision of Dr. Elham Semsar Kazerooni.

As a result of the work carried out over the last years, both in the development of this thesis and in the projects in which the doctoral candidate has participated, several scientific papers have been published in journals and congresses related to automated driving. Moreover, live demonstrations and other dissemination activities have been carried out in press and media. They are listed below.

Demonstration at IROS 2018

- "Autonomous Driving Events" at *2018 IEEE/RSJ International Conference on Intelligent Robots and Systems*(IROS 2018). A Link: https://youtu.be/jt4yudeznbw

Demonstration at S-Moving 2018

- Live demonstration. Canal Sur, Andalucéa, Spain, 18th October 2018, Link: https://youtu.be/Q9Oskis7iR0
- TVE News, Spain, 18th October 2018, Link: https://youtu.be/kgNnyXHkAUI

Demonstration at ICVES 2018

- INSIA facilities, 14th September 2018, Link: https://www.icves2018.org/uploads/5/0/3/9/50391481/icves_2018_demo.pdf

SEAT Autonomous Driving Challenge 2018

- "La Universidad Politécnica de Madrid gana la competición de coches autónomos a escala", ABC, 27th November 2018, Link: https://www.abc.es/motor/reportajes/abci-universidad-politecnica-madrid-gana-competicion-coches-autonomos-escala-201811270257_noticia.htmlhttps://www.abc.es/motor/...
- "Autopia: el proyecto que quiere resolver las dudas de la conducción autónoma", Innovaspain, 10th December 2018. Link: https://www.innovaspain.com/upm-universidad-politecnica-madrid-autopia-conduccion autonoma/
- "La UPM, campeona nacional del Autonomous Driving Challenge", Autocasión, 26th November 2018, Link: https://www.autocasion.com/actualidad/noticias/la-upm-campeona-nacional-del-autonomous-driving-challenge
- "La Universidad Politécnica de Madrid, campeona nacional del Autonomous Driving Challenge", RRHH press, 25th November 2018, Link: https://www.rrhhpress.com/talento/44374-la-universidad-politecnica-de-madrid-campeona-nacional-del-autonomous-driving-challengehttps://www.rrhhpress.com/talento/...
- "Estudiantes de la Escuela campeones del Autonomous Driving Challenge 2018.", Escuela de Industriales UPM, 30th November 2018, Link: https://www.escuelaindustrialesupm.com/ingeniero-industrial/estudiantes-de-la-escuela-campeones-del-autonomous-driving-challenge-2018/https://www.escuelaindustrialesupm.com/...

Dissemination Event: "Inventos Y Avances Científicos que están cambiando el mundo"

- "Comienza un ciclo de conferencias sobre los inventos que cambiarán el mundo", CSIC, Ciencia y sociedad, 10th January 2018, Link: http://www.csic.es/home?p_p_id=contentviewerservice_WAR_alfresco_packportlet&p_p_lifecycle=1&p_p_state=maximized&p_p_mode=view&p_p_col_id=column-1-3&p_p_col_count=2&_contentviewerservice_WAR_alfresco_packportlet_struts_action=
- "Ibercaja Huesca acoge el ciclo de conferencias" inventos y avances científicos que están cambiando el mundo", Web Fundación Ibercaja, 11th January 2018, Link: https://www.fundacionibercaja.es/salaprensa/7159
- "Los inventos que cambiarán el mundo", Catalunya Vanguardista, 11th January 2018, Link: https://www.catalunyavanguardista.com/los-inventos-que-cambiaran-el-mundo/

Author Biography

Dr. Antonio Artuñedo is currently a postdoctoral researcher in the AUTOPIA group at the Centre for Automation and Robotics (CSIC-UPM) in Madrid, Spain.

He received a B.Sc. in Electrical Engineering from the Universidad de Castilla—La Mancha, Spain in 2011 and a M.Sc. in Industrial Engineering from the Universidad Carlos III de Madrid in 2014. In 2019, he received his Ph.D. in Automation and Robotics at the Technical University of Madrid (UPM), Spain in the AUTOPIA Program. His Ph.D. degree was awarded with the "Cum Laude" distinction and the International Mention. During his predoctoral period, he made a research stay at the Integrated Vehicle Safety group at TNO, Netherlands, in 2017.

He joined the Centre for Automation and Robotics (CSIC-UPM) in 2013, where he has been working on both national and European research projects in the scope of autonomous vehicles. Antonio has published and peer-reviewed multiple journal and conference articles focused on this research field. His research interests include system modelling and simulation, intelligent control, motion planning and decision-making systems.

A. Artuñedo, *Decision-making Strategies for Automated Driving in Urban Environments*, Springer Theses, https://doi.org/10.1007/978-3-030-45905-5